ESG Integration in the Banking Sector

Elisa Menicucci

ESG Integration in the Banking Sector

Navigating Regulatory Frameworks and Strategic
Challenges for Financial Institutions

Elisa Menicucci
University of Roma Tre
Rome, Italy

ISBN 978-3-031-81676-5   ISBN 978-3-031-81677-2   (eBook)
https://doi.org/10.1007/978-3-031-81677-2

This Palgrave Macmillan imprint is published by the registered company Springer Nature Switzerland AG.
The registered company address is: Gewerbestrasse 11, 6330 Cham, Switzerland

If disposing of this product, please recycle the paper.

COMPETING INTERESTS

The author has no competing interests to declare that are relevant to the content of this manuscript.

Contents

LIST OF FIGURES

LIST OF TABLES

CHAPTER 1

Introduction

Abstract The environmental, social, and governance (ESG) dimensions have attained a dominant role for financial institutions in recent years and in particular the practical initiatives to incorporate ESG into business and operating models have become one of the most significant issues for European banks. The three letters E, S, G—which form the acronym ESG—identify a comprehensive setting that have changed and will continue to have a huge influence on banking. Why ESG topics matter for financial institutions and banks? What impacts do ESG dimensions have on banks? The significance of ESG matters has never been more prominent since the society has to deal with environmental and social challenges such as climate change, social inequality, and governance failures. In this context, ESG is increasingly becoming a relevant issue for doing business in banks as it can support in managing risks, enhancing reputation, complying with regulatory requirements, driving innovation, and growing access to capital. Banks that prioritize ESG are likely to attain long-term success because they are better appropriate for creating a more sustainable and fair future for all stakeholders. ESG themes are progressively important for customers, investors, and other stakeholders who want to recognize banks as responsible entities. Banks that prove their commitment toward sustainability are more able to fascinate customers, as well as to raise funding from ESG-interested investors.

© The Author(s), under exclusive license to Springer Nature Switzerland AG 2025
E. Menicucci, *ESG Integration in the Banking Sector*,
https://doi.org/10.1007/978-3-031-81677-2_1

1

Keywords ESG • Environmental • Social • Governance • Banks

In recent years, macroeconomic dynamics, social transformations, and regulatory and technological developments have determined significant changes in the banking sector, with significant consequences also for its future and developing trend. Considering the profound transformation of the global environment, banks and financial institutions are progressively challenged to incorporate environmental and social issues into their strategic planning to go beyond financial performance.

The environmental, social, and governance (ESG) dimensions have assumed a central role for financial institutions in recent years and in particular the practical initiatives to integrate ESG into business and operating models have become one of the most important topics for European banks. The three letters E, S, G—which form the acronym ESG—cover a comprehensive setting that have changed and will continue to have a massive impact on banking.

Why ESG issues matter for financial institutions and banks? What effects do ESG dimensions have on banks? It is shared that sustainability purposes and specifically the environmental, social, and governance (ESG) dimensions have already been increasingly incorporated in capital investments recently, and especially the COVID-19 pandemic period has intensified this trend. Hence, ESG factors identify an umbrella concept that should drive the investment choices. In this regard, the ESG investing offers even conventional investors an opportunity to foster sustainable businesses through their investments. Arguably, when numerous investors incorporate ESG dimensions in their investment choices, and thus directly and indirectly, they affect investee companies, and achieving sustainable development is clearly facilitated. The novelty compared to the past is also the priority that ESG issues have assumed in the global political agenda and, consequently, the major awareness of all stakeholders on the ESG concerns' centrality in ensuring the sustainability of business models of the intermediaries. Thus, the commitment toward sustainable business represents both a regulatory compliance obligation within a constantly evolving framework and a value creation opportunity in line with the double materiality paradigm.

The relevance of ESG considerations has never been more prominent since the society has to deal with environmental and social challenges such as climate change, social inequality, and governance failures. In this context, ESG is gradually becoming a significant issue for doing business in banks as it can aid in managing risks, enhancing reputation, complying with regulatory requirements, driving innovation, and growing access to capital. Banks that prioritize ESG are likely to achieve long-term success because they are better suited for creating a more sustainable and fair future for all stakeholders.

ESG issues are increasingly important for customers, investors, and other stakeholders who want to recognize banks as responsible entities. Banks that demonstrate their commitment toward sustainability are more able to attract and maintain customers, as well as to access funding from ESG-interested investors. For example, a bank that finances renewable energy projects or promotes social programs in its local community is probably positioned more positively than a bank that does not materialize ESG dimensions. For these purposes, banks must review their ESG strategies and their impact on profitability and risk to confirm their role in society in the perspective of sustainable finance. Then, banks should start to assess the implications that ESG factors may have on their core operations and business model in order to identify and prioritize the crucial functions to be revised (e.g. compliance, audit, product governance, risk management, risk controls, investment processes, etc.) in line with the already publicly available regulatory guidelines and market trends on ESG.

The growing relevance of ESG factors in the global economy has inevitable implications on the European banking industry. While EU authorities are designing the new ESG regulatory framework, banks are already addressing demands by their stakeholders (institutional investors, rating agencies, clients, etc.) for more real ESG solutions. Embedding ESG metrics into banking processes and testing their applications—for example, in product design, pricing and sales decisions, risk management, compliance functions—has become a sustainability challenge for banks and a critical topic for fostering bank revenues. Besides, regulators and authorities in the banking sector are deeply interested in this theme and they issued new requirements to foster the integration of ESG factors and risks in the banking system. The effective strategies of banks should include ESG dimensions into bank risk frameworks considering that ESG risk is not a fully stand-alone risk type. Based on a holistic approach to banking risks within the risk management process, ESG risks influence financial and non-financial banking risks through complex cause-effect-relationships.

This book explores ESG issues in the banking sector and it is expected to spark interest among stakeholders who are interested to the substance and the significance of the ESG dimensions, for example, companies, government, regulators, financial institutions, rating agencies, academics, practitioners, financial operators, etc. Specifically, this book provides a useful deep dive on these topics and offers a guidance for further development of practical application and initiatives on ESG concepts in the banking sector.

The book is structured as follows. The second chapter provides an overview of the ESG Regulatory Framework and describes the ESG framework as well as its pillars. Chapter 3 deepens the practical implementation of the ESG framework in the banking sector. Additionally, the chapter discusses the ESG challenges for banks and the real effect of integrating ESG dimensions in the European banks' operations. This chapter focuses on how banks can incorporate ESG dimensions into strategies and decision-making processes. Specifically, the investigation regards the impacts on banking processes, investment strategies, and client engagement.

The fourth chapter discusses the ESG risks and their assessment in banking strategies. The analysis highlights the necessary modalities for risk management and credit risk allocation according to the ESG metrics and illustrates how ESG risks can affect financial and non-financial risks in banks. Finally, the last chapter considers the concept of double materiality applied to ESG issues in the banking sector. According to this perspective, both the ESG-related impacts on the bank and the impacts of a bank on the ESG dimensions can be material. The chapter describes the double materiality assessment, its rationales, challenges and strategic implications for banks and underlines the role of double materiality in implementing and realizing the ESG strategy in banking.

CHAPTER 2

ESG Regulations and Concepts

Abstract This chapter provides an overview of the European regulatory scenario concerning the ESG issues and it describes the change of banking regulation toward ESG challenges. The context is characterized by a strong evolution which has expected and will expect important innovations in terms of climate and environmental risks in the financial sector. The main European authorities have established main sustainability objectives and requirements and they have foreseen some roadmaps to promote the transition toward sustainable finance by integrating ESG factors.

The European Banking Authority (EBA) published in December 2019 an Action Plan for progressive integration of ESG factors in the players' business strategies market through the identification of targeted actions to identify, measure, and monitor such risks and support the path toward sustainable finance. Also, the European Central Bank (ECB) has prepared a roadmap aimed at the progressive integration of ESG factors from the banking system. In November 2020 the ECB and the Authorities Competent Nationals (ANC) developed jointly the "Guidance on environmental climate risks" which reports the view of the Supervisors regarding climate-environmental risk management. The Guide also describes the expectations of how to take ESG issues into account in formulating and implementing corporate risk strategies. This Guide is not mandatory for banks, but it is relevant for the significant banks subjected to the direct supervision of ECB.

E. Menicucci, *ESG Integration in the Banking Sector*, https://doi.org/10.1007/978-3-031-81677-2_2

The European Regulatory Framework on the topic is made up of EU Regulations with the aim of directing investments toward sustainable projects and activities. In this regard, the Proposal for a Commission Directive European Union 2021/189 of 04/21/21 (Directive CSRD), was issued with the dual objective of integrating previous Directives about corporate communication on sustainability and explaining the information set in more details, in compliance with EU principles on sustainability information. A key element of the EU CSRD is the setting of the European Sustainability Reporting Standards (ESRS) to which companies under the CSRD must align their reporting from 2024. The aim of these Standards is to ensure that in-scope companies report comparable, relevant, and reliable sustainability information as well as to make clear what they are expected to report on.

Keywords ESG regulations • CSRD • ESRS • ESG pillars

2.1   INTRODUCTION

The environmental, social, and governance (ESG) dimensions have assumed a central role in recent years and they have attracted great attention from European banking authorities. The banking sector has gained a growing awareness of the need to integrate ESG factors into corporate strategies and business models in the perspective of sustainable finance. This objective represents not only a regulatory compliance obligation deriving from a constantly evolving regulatory framework, but above all it is a medium-long term value creation opportunity for banks. ESG dimensions provide an umbrella framework to consider a bank's impact on the environment and society and on the quality of its corporate governance. In this regard, the consideration of ESG factors alongside financial factors in the investment decision-making process may have a positive or negative impact on the financial performance of a financial institution. Hence, we describe the ESG framework and offer an in-depth analysis of the three ESG pillars (i.e. environmental (E), social (S), and governance (G)) as well as the interactions between them. We discuss the theoretical basis of ESG Framework, and we explore the concept of ESG by establishing a single definition of each ESG pillar to facilitate the understanding in a consistent way. We also provide an overview of the main sustainability objectives established at the level as well as the roadmap and requirements established by the main European authorities for the sector finance to promote integration of ESG factors and risks and the transition toward sustainable finance.

2.2   The ESG European Regulatory Scenario

Regulators are progressively converging on environmental, social, and governance (ESG) issues and requiring banks to integrate these themes into their business practices. Financial markets have seen a progress in ESG-focused shareholder activism as an active ownership strategy.[1] Such an expansion of investor engagement in ESG issues has been attributed to the growing public interest of institutional investors in ESG topics.[2] The financial system is now on the eve of a new phase which will lead to a significant increase of the ESG information set which will be offered to stakeholders in the coming years. In this context, banks play a central role in providing credit and financial resources that can be used to alleviate the negative effects of ESG risks and to allow the economy in becoming more resilient in this regard. Banks are facing new challenges and in particular an appropriate consideration of ESG risks in a wide range of bank processes will be crucial for increasing bank profitability. Besides embedding ESG into risk frameworks, banks need to consider related issues in product design, pricing, and sales decisions. For these reasons, banks currently need a substantial commitment of managers and consultants and a correlated considerable investment in human resources, processes, and activities for the definition of the new system of planning, reporting, and communication of non-financial information to achieve the underlying sustainability objectives.

The inclusion of ESG factors in bank operations and strategies has been gradually attained through the performing of a heterogeneous package of legislative and regulations issued by the European Legislators. In recent years, European Authorities have strengthened their efforts to produce a

---

[1] See Eccles, G.R., Klimenko, S. (2019), "The Investor Revolution", *Harvard Business Review*, Vol. 97, pp. 106–116.

[2] See Hoepner, A.G.F., Oikonomou, I., Sautner, Z., Starks, L.T., Zhou, X. (2021), "ESG Shareholder Engagement and Downside Risk", ECGI Finance Working Paper No. 671/2020, European Corporate Governance Institute, Brussels, Belgium.

regulatory framework that encompasses ESG principles in the banking sector for improving sustainability in the financial system. The regulatory scenario is characterized by a strong evolution which has provided and will expect to run significant changes in terms of ESG risks in the financial sector for a more sustainable economic system. According to the European Banking Authority (EBA), it is necessary to incorporate ESG risks into the banking regulatory and supervisory framework, giving a particular emphasis on climate and environmental risks due to the climate change and the governments' requirements to move toward a "green economy". Since ESG risks may have different impacts on banks' financial performance, the supervisory process is changing in line with these new risks and, in particular, banks are obliged to confirm their capacity and ability to manage adequately the effect of these risks within the assessment of the viability and sustainability of banks' business model. For this reason, supervisors are interested in the analyses implemented by the banks and in non-financial reporting that contains a number of information useful to discover the level of attention to a sustainable strategy and ESG ratings. How do these new types of ESG risks affect different banking activity? The main international and European Regulatory Authorities assessed the importance of ESG risks for banking activities considering that banks will need to reassess their business strategies and to reinforce their risk management frameworks according to the new perspectives of ESG risks.[3] In any case, as risks are global and systemic, a regulatory coordination on ESG risks is necessary. In this regard, a recent ECB (European Central Banks) assessment shows that banks have made some operations progress in conforming their practices to manage these risks and to meet the supervisory expectations.[4] In this regard, Supervisors have already scheduled some specific measures

---

[3] See Alogoskoufis, S., Nepomuk, D., Emambakhsh, T., Hennig, T., Kaijser, M., Kouratzoglou, C., Muñoz, M.A., Parisi, L., Salleo, C. (2021), "ECB's economy-wide climate stress test," Occasional Paper Series 281, European Central Bank; Elderson, F. (2021), "Overcoming the tragedy of the horizon: requiring banks to translate 2050 targets into milestones", speech at the Austrian Financial Market Authority's Supervisory Conference, October 20.

[4] See ECB Banking Supervision, (2021), "The state of climate and environmental risk management in the banking sector. Report on the supervisory review of banks' approaches to manage climate and environmental risks", November.

for next years and beyond,[5] including a thematic review of banks' environmental risk management practices and a stress test on climate-related risks.

During the last years, the financial sector has been involved in a great business development, which has been characterized by the introduction of the new ESG principles. These principles are driving banks toward an innovative idea of management that incorporates a wide interpretation of the concepts underlying the ESG principles. In general, banks are becoming more and more active in investment and asset allocation as well as in new business models with the final goal of a more sustainable framework for financial activity with a selection of assets and sectors to finance. The attention to the environment and its exploitation and to the reduction of pollution or carbon emissions are influencing the banks' choices and strategies, and also a new attention to social justice and social principles is becoming very relevant. Hence, regulators, rating agencies, and other parties are taking a keen focus on these topics, leading to increasing requirements and reporting needs for financial institutions. For this reasons, the European regulatory framework for sustainable finance has greatly developed and this constant flow of new regulations is leading to extensive compliance challenges for banks about ESG principles. In this context, banks play a very significant role as they have the power to lead this new challenge and are able to facilitate businesses to run toward a sustainable green economy. Hence, banks' activity is now oriented to rise and allocate credit and investment to more sustainable sectors and businesses. As climate risk is both a cause and an effect for socially responsible banking operations, climate change has shown the importance of climate risks and their significant implications in the new risk management process and Regulators have considered the role of banks for the green and ecological transition.

In this chapter, we present some reflections on the evolution of European policies on sustainability issues for financial intermediaries supporting the idea of a necessary strengthening of the awareness on these aspects in the future. In line with the academic and institutional debate on ESG in the banking sector, we support the need for improving the awareness about ESG issues, sharing the view that the financial system can act as a catalyst agent for the transition to a more sustainable economy as a

---

[5] Many of the proposed regulatory changes actually derive from research conducted by the EBA and the ECB on the use of internal models applied by European banks.

whole. We want to contribute to the on-going debate on ESG especially in the banking sector by offering a systemic overview of the European policies regarding ESG issues.

The European Union, in recent years, has started a series of reforms with the aim of achieving an advantage in the context of the sustainable transition compared with the rest of the world. In this regard, European Authorities are working toward the development of a progressively sustainable economic-financial system, aimed at making firms aware and responsible of its impact on the earth. This is in line with the European commitment to climate neutrality to be achieved by 2050, with the structural measures linked to the Industrial Green Deal and with the presentation of the roadmap for Sustainable Finance by the EBA (European Banking Authority). For some time now, large firms have begun to disclose non-financial information related to the impact of corporate operations on the environment and society. In particular, the banks, especially the most relevant ones, have proceeded to draw up accurate declarations and report on a wide range of data and information (e.g. energy transition activities, the amount of environmental financing, ESG investment products, sustainable bonds and social finance, the integration of ESG factors in business), initially simply as a challenge and then as a new requirement. For these reasons, the European Union has introduced regulations which require banks to align their investments with environmental objectives and to disclose ESG-related information. This is a process which, on the one hand, strengthened the awareness and the knowledge of the topic, on the other hand, it required the arrangement of a reorganization work after an initial period of application. The main efforts have been assigned to the reinforcement of the disclosure requirements for banks, financial intermediaries, and industrial companies.[6] The main regulations that are worth mentioning are the following.

The turning point is the Legislative Decree 254/2016, which implemented the Directive 2014/95/EU of the European Parliament and of the Council of October 22, 2014 (amending the Directive 2013/34/EU). This is a European Union (EU) Directive on Non-Financial Reporting (NFRD) intended for public interest entities (banks, insurance companies, listed companies, and those operating on the capital market)

---

[6] Overall, all the reporting initiatives can be considered beneficial since the disclosure requirements are quite diverging nationally.

of large dimensions, introducing the obligation to prepare the Non-Financial Declaration (NFD) to integrate ESG factors within traditional financial reporting. We also refer to the NFRD issued in 2014, the Sustainable Finance Disclosure Regulation (SFDR), and the European Commission Guidelines on non-financial reporting (the "Guidelines") of 2019, the Taxonomy Regulation of 2020, and the EU Directive no. 2022/2464 regarding Corporate Sustainability Reporting Directive (CSRD) issued on 16 December 2022 within the scope of the European Green Deal. The CSRD modified the Directive 2013/34/EU (NFRD), concerning the obligation to communicate information of non-financial nature for large companies, with the aim of strengthening and standardizing the disclosure obligations and limiting the phenomenon of greenwashing.

Starting from 2025, the European Union will require large and listed companies (except listed micro-enterprises) to disclose information on risks and opportunities related to their ESG practices, with a particular focus on the impact of their activities on people and the environment. The Corporate Sustainability Reporting Directive (CSRD) replaces the EU's legacy ESG reporting program—the NFRD—and calls for a greater breadth and soundness in sustainability reporting, including disclosure about carbon, pollution, water, waste, and biodiversity. The CSRD's technical rules known as the European Sustainability Reporting Standards (ESRS) set up what companies need to disclose and how on these topics in annual reports alongside financials.

Additionally, the EU Regulation no. 2020/852, so-called "Taxonomy Regulation", introduced the methodology to be applied in order to classify financial investments as eco-sustainable, where they comply with certain requirements and characteristics. In other words, this Regulation aims to define a common language and a definition of "sustainable activities".

This unified system of classification of sustainable economic activities in Europe (i.e. the Taxonomy) was established with EU Regulation 2020/852 and aims to encourage investments with environmental and social objectives. The Taxonomy defines six environmental objectives, which are consistent with the minimum environmental reporting requirements envisioned by the CSRD: mitigation of climate change, adaptation to climate change, sustainable use and protection of water and marine resources, transition toward a circular economy, pollution prevention and control, protection and restoration of biodiversity and ecosystems. As required by EU Regulation 2020/852, starting from January 1, 2023, the companies that are subject to the obligation of publishing non-financial information must also report information regarding the economic activities both aligned and not with the EU Taxonomy (i.e. the Taxonomy alignment). The EU Taxonomy therefore aims to provide appropriate definitions for which economic activities can be considered sustainable from an environmental point of view, encouraging sustainable investments, preserving the private investors from greenwashing, supporting companies in developing greater sensitivity with respect to climate-related issues, mitigating the market fragmentation, as well as favoring the allocation of investments. Finally, beyond these objectives and potential benefits illustrated, the EU Taxonomy has the ultimate aim of promoting the sustainable investments and implementation of the European Green Deal. In addition to the Taxonomy Regulation, the European Commission has developed the list of environmentally sustainable activities defining technical screening criteria for each objective through delegated acts. Finally, with the Delegated Regulation (EU) n. 2021/2178 (the "Delegated Regulation"), which combined the Taxonomy Regulation, the content and method of presentation of the information that companies subject to the Non-Financial Declaration (NFS), pursuant to the NFRD Directive, must communicate about eco-sustainable economic activities. The Delegated Regulation specified the key performance indicators (KPIs) through which the information must be disclosed, as well as the related calculation methodologies. These KPIs have been defined differently, depending on whether they are addressed to the financial companies, non-financial companies, investment companies, banks, insurance, and reinsurance companies.[7]

---

[7] The main of these KPIs is the Green Asset Ratio (GAR)—understood as the ratio between credit assets that finance economic activities aligned with the Taxonomy (numerator) and total assets (denominator)—aimed at including information relating to the Taxonomy. In March 2021, the EBA underlined the importance of the green asset ratio (GAR) as a key means to understand how institutions are financing sustainable activities and meeting the Paris agreement targets. See EBA, Call for Advice to the European Supervisory Authorities on Key Performance Indicators and Methodology on the Disclosure of How and to What Extent the Activities of Undertakings under the NFRD Qualify as Environmentally Sustainable as Per the EU Taxonomy. EBA-2020-D-3408, February 26, 2021.

2.3   THE EUROPEAN BANKING AUTHORITY (EBA)'S ACTION PLAN ON ESG

In the context described above, the main European Authorities have foreseen some roadmaps focused on encouraging the transition toward sustainable finance by integrating ESG factors. Some recommendations and regulatory developments were issued regarding the need for early and proactive actions to ensure readiness for ESG-related challenges. Between these, the European Banking Authority (EBA) published in December 2019 an Action Plan for a progressive integration of ESG factors in the players' business strategies through the identification, measurement, and monitoring of ESG risks to pave the way toward sustainable finance. In particular, regarding the integration of ESG risks, the EBA highlights the importance of proper identification, management, and monitoring of them as they may lead to potential impacts to the traditional risks, such as credit, operational, and market risks.

To include ESG risks into the three pillars of the banking prudential framework, EBA proposes an integration approach composed by three potential phases: (1) assessment of the exposure to ESG risks; (2) evaluation of the exposure to ESG risks; (3) undertaking of adaptation initiatives.

Regarding business strategies, considering that banks are, and should continue to be, responsible for setting their strategies, the impacts of ESG risks should be properly considered in order to guarantee the resilience of business models over the short-, medium-, and long-term time horizons. The EBA recommends that the banks achieve this by the incorporation of ESG risks into banks' business strategies, into internal governance arrangements, and into risk management frameworks according to a risk-based and proportionate approach. In particular:

- Including ESG risk-related considerations when setting business strategies, in particular by spreading the time horizon for strategic planning to at least 10 years, at least qualitatively, and by testing their resilience to different scenarios.
- Setting, disclosing, and implementing ESG risk-related strategic objectives and/or limits, counting related key performance indicators (KPIs), in accordance with the bank's risk appetite.
- Engaging with counterparties and clients, borrowers, investee companies, and other stakeholders.

- Assessing the potential necessity to develop sustainable products or to correct features of existing products, as a way to contribute to and safeguard the alignment with strategic objectives and/or limits.

Regarding Risk Management, the EBA recommends that banks integrate ESG risks into their risk management framework, considering the assessment of their materiality over different time horizons, by:

- Embedding material ESG risks in the risk appetite framework.
- Managing ESG risks as drivers of financial risks, in line with the risk appetite framework (RAF) and as it is reflected in both the Internal Capital Adequacy Assessment Process (ICAAP) and Internal Liquidity Adequacy Assessment Process (ILAAP) frameworks.
- Identifying the gaps the banks are facing in terms of data and adequate methodologies and take the consequence remedial actions.
- Setting out proper policies considering ESG for the assessment of the financial robustness of counterparties.
- Developing risk monitoring metrics at exposure, counterparty, and portfolio level.
- Developing methodologies to test the bank resilience to ESG risks, understanding the robustness of the bank's business model and investment strategies.

Regarding corporate governance, the EBA recommends that banks incorporate ESG risks in governance structures, implementing operating procedures and responsibilities for business lines and internal control functions to assure a complete and comprehensive approach for the incorporation of ESG risks into business strategy, business processes, and risk management. This should regard the management body and its "tone from the top" approach concerning risk culture, allocation of tasks, and responsibilities related to ESG risks as drivers of financial risk categories in the decision-making process and arrangements for an effective management of ESG risks.

In addition to the EBA Action Plan, the European Central Bank (ECB) has prepared a roadmap aimed at the progressive integration of ESG factors in the banking system to adapt the supervisory and evaluation process (SREP). In such scope, in November 2020 the "Guidance on environmental climate risks" (or "Guide"), developed jointly by the ECB and the Authorities Competent Nationals (ANC), reported the view of the

Supervisor Authorities regarding risk management climate-environmental issues and described expectations on how to take these risks into account in formulating and implementing corporate strategies, governance, and risk management systems. This Guide is not binding for banks but it will be relevant in the context of the relationships with the ECB and the significant banks subjected to the direct supervision of the ECB.

The Guide indicates the Supervisor's expectations concerning the following main areas:

- Business models and business strategy, which entails the identification of the climate and environmental risks in relation to key sectors and geographical areas as well as to products and services in the short, medium, and long term to be compliant with a dedicated sustainability strategy.
- Risk management framework, which provides the integration of climate-environmental risks within the existing risk management system, introducing risk limits and metrics within the RAF until the qualitative-quantitative adjustment of the ICAAP and ILAAP frameworks. The inclusion of climate and environmental risks in the risk management system may be carried out, for example, through the definition of risk limits of climate-environmental measures settled on the basis of the principle of proportionality, with the aim of achieving effective management of this category of risks in accordance with the bank's "as is" processes management, monitoring, and reporting. Overall, this integration will have to be realized ensuring synergy and coherence with business strategy and management strategy of the bank.
- Internal governance, which provides the inclusion of climate and environmental issues in the corporate strategy. With regard to the internal governance, the Guide refers to dedicated adaptation interventions, such as the attribution of specific roles and skills regarding ESG issues to existing committees or the creation of new committees ad hoc for the development of critical skills on ESG topics. These issues related to monitoring, reporting, and disclosure must be conveyed to top management to assess the performance trend from an ESG perspective and its possible development areas and initiatives.
- Disclosure, which integrates the current disclosure processes through qualitative-quantitative information regarding the climatic and environmental aspects.

2.4   The Corporate Sustainability Reporting Directive (CSRD) and the New Mandatory European Sustainability Reporting Standards (ESRS)

The European Regulatory Framework on the topic of ESG is made up of EU Regulations with the aim of directing investments toward sustainable projects and activities. In this regard, the EU Directive no. 2022/2464 (Directive CSRD),[8] issued with the aim of integrating previous Directives about corporate communication on sustainability, explains the information set in more detail for communication on sustainability, in compliance with EU principles on sustainability information. This Directive extends the disclosure obligation requiring the certification of information on sustainability. The new rules will ensure that both investors and other stakeholders are informed to ascertain the impact of banks on the environment and for investors to assess financial risks and opportunities resulting from climate change and other sustainability issues. The Directive CSRD aims to promote transparency by improving the reliability and the comparability of sustainability information reported throughout the EU. The CSRD provides for progressive application, starting from January 1, 2024, for companies already subjected to the reporting obligations of the Non-Financial Reporting Directive (Direttiva 2014/95/EU, "NFRD").[9]

A key element of the CSRD is the set of the European Sustainability Reporting Standards (ESRS) that companies under the CSRD must align their reporting to from 2024. The new EU regulation integrates the ESG reporting directive by setting the new sustainability reporting principles,

---

[8] On January 5, 2023, the Corporate Sustainability Reporting Directive (CSRD) entered into force. It updates and improves the rules concerning the social and environmental information that companies have to report. A broader set of large companies, as well as listed SMEs, will now be required to report on sustainability. Some non-EU companies will also have disclosure on these issues if they generate over EUR 150 million on the EU market.

[9] This group of companies includes public interest entities such as banks, insurance companies, and listed companies that exceed certain size limits. The first companies will have to apply the new rules for the first time in the 2024 financial year, for reports published in 2025. Starting from 2025, all large corporations that meet specific criteria related to employees, total assets, and sales will be required to report. This includes both listed and non-listed enterprises. However, the Standards for Small and Medium-sized Enterprises (SMEs) are yet to be defined, and they may only be required to report starting from 2027. With the purpose of guaranteeing the application of the Directive, Member States has been delegated to provide a system of "effective, proportionate and dissuasive" sanctions for the violation of national provisions in application of the Directive.

called European Sustainability Reporting Standard (ESRS). The ESRS was developed by the European Financial Advisory Group (EFRAG) and adopted as a delegated act by the European Commission on July 31, 2023. Since January 2024, the European Sustainability Reporting Standards (ESRS) have set the framework for sustainability reporting in Europe.

By requiring the use of common standards (ESRS), the EU directive aims to ensure that companies across the EU report comparable and reliable sustainability information and to make a greater commitment in assessing the risks and their impacts on the sustainability of their business. Enforcing the first 12 European ESRS,[10] the CSRD aims to standardize and enhance the transparency of ESG reporting across the EU by applying common standards that will help firms to reduce reporting costs in the medium and long term and to avoid the use of multiple voluntary ones. Hence, the CSRD assures that in-scope companies report comparable, relevant, high-quality, and reliable sustainability information to create a culture of greater public accountability.

The development of these ESRS principles is based on two elements: the alignment with legal EU frameworks and the consistency with existing reporting recommendations and standards. Firstly, to create a unified reporting landscape in the EU that supports the goals of the EU Green New Deal and Sustainable Finance Framework, the ESRS reporting areas are coherent with other EU frameworks and legislations such as the EU Taxonomy and Sustainable Finance Disclosure Regulation (SFDR). Secondly, to guarantee a high level of interoperability between the ESRS and other international standards, the ESRS reporting requirements are aligned with the recommendations of the Task Force on Climate-Related Financial Disclosures (TCFD) and the standard of the Global Reporting Initiative (GRI).[11]

---

[10] The first set is made up of 12 Standards: two cross cutting Standards of a general scope and ten topical Standards (Environmental, Social, Governance) divided by topic: 5 environmental issues, 4 social topics, and 1 on governance. The CSRD Directive also requires EFRAG to continue its work with the further issuing of sector-specific Standards.

[11] The Standards are designed to be highly interrelated with the GRI Standards, to be consistent with the recommendations of the Financial Stability Board's TCFD (Task Force on Climate Related Financial Disclosures), and to reflect the disclosure requirements issued by the EU Green Taxonomy and the Corporate Sustainability Directive Due Diligence (CSDD).

The transition to ESRS reporting under the CSRD represents both a challenge and an opportunity for EU companies. By adopting a proactive approach to understanding and implementing the ESRS requirements, companies can improve their sustainability practices, enhance their corporate transparency, and contribute positively to the global sustainability agenda. For successful reporting under the ESRS, companies must engage in advanced preparation and efficient planning. This approach involves a comprehensive comprehension of the extensive scope of the ESRS, identifying any gaps in existing data and reporting processes, and planning the workflows and the necessary information for compliant reporting. This strategic planning is necessary to conform with the ESRS's requirements and to demonstrate leadership in sustainability just meeting regulatory disclosure demands. The ESRS framework aims to provide a transparent, accurate, and comparable overview of a company's ESG impacts, risks, and opportunities. This is not simply a regulatory requirement but a move toward incorporating sustainability into the heart of business reporting, reflecting the European Union's commitment to a sustainable economic future. By exacting detailed disclosure, the ESRS encourage European companies to reassess their sustainability practices and align them with global best practices, setting a new benchmark for corporate accountability and transparency to realize a more informed and sustainable marketplace for all stakeholders.

What are the milestones that characterize and distinguish these principles? The ESRS take a "double materiality" perspective. According to this approach, the standards require companies to report their impacts on society and the environment and how social and environmental issues generate financial risks and opportunities for the company. According to the ESRS and double materiality assessment, a sustainability matter is material when it meets the criteria defined for the materiality of the impact or for financial materiality or for both. In this regard, companies have to give sustainability information regarding both the impact of its business activities on society and the environment (inside-out approach) and the way in which sustainability issues influence them and their results (outside-in approach). Banks will have to define sustainability strategies and to integrate ESG objectives into their strategy: it will be appropriate to include information necessary to understand how sustainability initiatives affect the bank performance, the economic-financial situation, as well as the structure of the business model. The ESRS provide a framework to incorporate the ESG aspects along the value chain and to include risks within the Enterprise Risk Management (ERM). The reporting of sustainability information has to include information on material impacts, risks, and opportunities related to the entire value chain upstream and downstream, as results of the activities of due diligence and materiality analysis. To respond to the changing nature of the risks to which banks are exposed and to the growing interest of investors on the deriving financial implications, banks will be obliged to consider, internally of the risk management model (ERM), the risks linked to environmental and social issues.

2.5   The ESG Framework

ESG is a framework used to assess an organization's business practices and performance on various sustainability and ethical issues. At its simplest, ESG provides an umbrella to consider a company's impact and dependencies on the environment and society, as well as the quality of its corporate governance. This reporting framework is also applied for the disclosure of data comprising business operations, opportunities, and risks that are related to the ESG aspects of the business. The ESG definition is rather linear but the challenge is to define the indicators identifying the ESG themes and to measure them. From another perspective of analysis, ESG can be considered an investment philosophy that considers economic, environmental, social, and governance benefits for generating a value of sustainable development. As an investment concept for evaluating the sustainable development of companies, the three basic pillars of ESG are the key points to be considered in the process of investment analysis and decision-making. In this sense, ESG is usually painting as a strategy used by investors to assess corporate sustainability behavior and future performance. A fundamental part of this sustainable development model is the evaluating and measuring of ESG risks uniquely to establish a common definition of ESG factors and to understand how these factors reflect into financial risks that in turn may impact financial institutions individually and the financial system as a whole.

ESG is also viewed as a fundamental strategy of corporate sustainability especially in the banking sector since it is a set of metrics for a bank's operations that socially and environmentally sensible investors use to appraise future investments.[12] Hence, effective strategic decisions to

---

[12] See Chen, Hong-Yi & Yang, Sharon S., 2020. "Do Investors exaggerate corporate ESG information? Evidence of the ESG momentum effect in the Taiwanese market," Pacific-Basin Finance Journal, Elsevier, vol. 63(C).

allocate resources and capital to improve the commitments relating to ESG help banks to increase customer loyalty and to achieve a sound financial position.[13] Certainly, banks must be cautious in investing in projects that are environmentally harmful as well as in engaging themselves in ESG controversies.[14] In this regard, we remember that ESG pillars stem from the concept of the responsible investment whereas the Principles for Responsible Investment (PRI) define a responsible investment as "a strategy and practice to incorporate environmental, social and governance (ESG) factors (or pillars) in investment decisions and active ownership".[15]

The ESG identifies a framework system including environmental (E), social (S), and governance (G) factors alongside financial factors in the investment decision-making process (see Table 2.1).[16] As the EBA (European Banking Authority) states, ESG factors are "environmental, social or governance matters that may have a positive or negative impact on the financial performance or solvency of an entity, sovereign or

---

[13] See Arli, D.I., Lasmono, H.K. (2010), "Consumers' perception of corporate social responsibility in a developing country", *International Journal of Consumer Studies*, Vol. 34 No. 1, pp. 46–51; Buallay, A. (2019), "Is sustainability reporting (ESG) associated with performance? Evidence from the European banking sector", *Management of Environmental Quality: An International Journal*, Vol. 30, No.1, pp. 98–115; Shakil, M.H., Mahmoos, N., Tasnia, M., Munim, Z.H. (2019), "Do environmental, social and governance performance affect the financial performance of banks? A cross-country study of emerging market banks", *Management of Environmental Quality: An International Journal*, Vol. 30 No. 6, pp. 1331–1344; Buallay, A. (2020), "Sustainability reporting and firm's performance: comparative study between manufacturing and banking sectors", *International Journal of Productivity and Performance Management*, Vol. 69 No. 3, pp. 431–445.

[14] ESG controversies include negative news about a bank's social and environmental scandals, lawsuits, and failure of corporate governance. ESG controversies may affect bank's financial performance, and adversely impact the market reputation. See Aouadi, A., Marsat, S. (2018), "Do ESG Controversies Matter for Firm Value? Evidence from International Data", *Journal of Business Ethics*, Vol. 151 No. 4, pp. 1027–1047.

[15] The six Principles for Responsible Investment (PRI) offer a range of possible actions for incorporating ESG issues into investment practice. They act to understand the investment implications of environmental, social, and governance (ESG) factors and to support its international network of investor signatories in considering these factors into their investment and ownership decisions. In implementing them, signatories contribute to developing a more sustainable global financial system.

[16] The source of the table is an own elaboration based on the EBA report on ESG risk management and supervision.

Table 2.1   The ESG framework

Dimension/ Pillar	Definition	Factors
Environmental	The environmental pillar assesses a company's impact on the environment, including its carbon footprint, waste management, resource conservation, and response to climate change. The pillar includes controls of: **carbon emissions**, impact on **deforestation and nature loss**, over-consumption of **non-renewable resources**, and production of **waste products.** It also includes positive contributions such as the **financing of environmental improvements** (e.g. green finance initiatives). The impacts of a company on the environment or the impacts of the environment on the company are both considered.	• GHG (greenhouse gases) emissions • Energy consumption and efficiency • Air pollutants • Water usage and recycling • Waste production and management (water, solid, hazardous) • Impact and dependence on biodiversity • Impact and dependence on ecosystems
Social	The social pillar assesses the company's treatment of employees, customers, and communities, as well as human rights, labor practices, and diversity and inclusion initiatives. This pillar evaluates the contribution of a company to fairness, equality, trust, and welfare in society (including the improvement of labor rights and diversity and inclusion) within a company's workforce, and across its supply chain and distribution.	• Workforce freedom of association • Child labor • Forced and compulsory labor • Workplace health and safety • Customer health and safety • Discrimination, diversity, and equal • Opportunity • Poverty and community impact • Supply chain management • Training and education • Customer privacy • Product safety • Data security • Community impacts
Governance	The governance pillar assesses a company's management and leadership, including the quality and the scope of reporting, corporate governance, corporate culture, transparency, ethical behaviors on both ESG and non-ESG matters, type of accountability, processes for decision-making, level of independent oversight, and ethical behaviors.	• Codes of conduct and business principles • Accountability • Transparency and disclosure • Executive compensation • Board diversity and structure • Director and audit independence • Bribery and corruption • Stakeholder engagement

individual".[17] From this perspective, ESG factors can be used also when entities evaluate opportunities related to the transition to a more sustainable economy. In this regard, the ESG factors support in measuring the sustainability and the social impact of business activities. Entities may be impacted to a varying degree by the transition to an environmentally sustainable economy. In fact, a comprehensive, long-term, and strategic approach to ESG factors and particularly the relevance of ESG factors for entities depend both on the business activities (e.g. asset type, sector, size, geographic location, and liabilities) and on the strategy and governance for managing them within an organization.

ESG factors show one or more of the intrinsic features presented in Table 2.1, which may possibly interrelate with each other. Moreover, ESG factors present the following commonalities:

- Are factors usually considered as non-financial and they embody characteristics such as greenhouse gas emissions, environmental footprint, social welfare, poverty, equal rights, and ethics, in addition to those factors that have been usually considered financial, such as profits, capital, and costs.
- Are characterized by uncertainty about their impact: these impacts may occur at any time (short, medium, and/or long term) and trigger effects over very different timespans. ESG factors are relevant in the medium and/or longer term, but they also generate risks in the short term, such as acute environmental hazards and the unexpected implementation of environmental policies.

---

[17] Cfr. EBA, EBA Report on management and supervision of ESG risks for credit institutions and investment firms, EBA/REP/2021/18, § 35. About social responsible practices and financial performance in banking, see also Simpson, W.G., Kohers, T. (2002), "The link between corporate social and financial performance: Evidence from the banking industry", *Journal of Business Ethics*, No. 35, No. 2, pp. 97–109; Esteban-Sánchez, P., de la Cuesta-González, M., Paredes-Gazquez, J.D. (2017), "Corporate social performance and its relation with corporate financial performance: International evidence in the banking industry", *Journal of Cleaner Production*, 2017, Vol. 162, pp. 1102–1110; Hernández, J.R., Quirós, M., Quirós, J.L. (2019), "ESG performance and value creation in the banking industry: International differences", *Sustainability*, Vol. 11, p. 1404; Brogi, M., Lagasio, V. (2019), "Environmental, social and governance and company profitability: Are financial intermediaries different?", Corporate Social Responsibility and Environmental Management, Vol. 26, pp. 576–587; Menicucci, E., Paolucci, G. (2023), "ESG dimensions and bank performance: an empirical investigation in Italy", Corporate Governance, Vol. 23 No. 3, pp. 563–586.

In Table 2.1, we present a framework for breaking down ESG into its structured and manageable sub-elements.

2.5.1   The ESG Pillars

ESG identifies a broad spectrum of topics that can affect all components of a company's strategy, business model, and operations. To make it easier to tackle the concept underlying the ESG framework, it would be appropriate to do a deep dive on the three pillars of ESG separately, and then on the sub-elements of E, S, and G pillars (dimensions) at the next level down. The three ESG pillars are closely interconnected, but each one has its own specificities.

The Environmental Pillar (E) describes how private and public entities interact with, affect, and respond to changes in the physical environment. There is a broad agreement that the environmental considerations usually refer to climate change, physical disasters, air and water pollution, resource exhaustion, and biodiversity loss. The environmental perspective can include the positive and negative environmental impacts and dependencies generated in the provision of goods and services, for example, damage for companies or citizens caused by extreme weather events or a decline of asset value of a company in carbon-intensive sectors. The issue of the environment is vital for the future of the planet and it involves a particularly large number of public and private players in the financial sector. Banks are starting to integrate the ESG dimensions into their business models, beginning from their internal operating and strategic processes. In this regard, there is a strong awareness of how much remains to be realized since a strong growth in demand for "sustainable" products from investors has already upsurged. Also at the urging of the supervisory authorities, the impact of environmental risks on financial activities is therefore under way to re-evaluate how these risks translate into the traditional ones (e.g. credit, market, and liquidity). Driven by environmental factors, climate-related risks are the financial risks derived by the exposure of companies to counterparties that may potentially contribute to or be affected by climate change or other forms of environmental degradation (e.g. air pollution, water pollution, scarcity of fresh water, land contamination, biodiversity loss, and deforestation). The Environmental Pillar also reflects how well a company uses the best management practices to avoid environmental risks (a company's impact on living and non-living natural systems, including the air, land, and water, as well as complete ecosystems) and capitalize on environmental opportunities to generate long-term shareholder value.

Environmental analysis considers entities' exposure to physical and transition risks and how they preserve or reduce natural capital. We identify the following environmental risks:

- Climate transition risk which embodies the implications of the world's transition to a low carbon and greener economy through a reduction in greenhouse gas emissions, shifts in public sentiment and consumer demand, changes in climate policy or legislation, or technological advancements which could lead to significant fluctuations in demand and high development costs of certain products and services.
- Physical risk which is associated with the environment and can be characterized by circumstances such as wildfires, floods, droughts, or hurricanes; or, exemplified by longer-term shifts in climate patterns such as increasing temperatures, rising sea levels, or increased weather variability. Physical risks also comprise other natural disasters that are not always climate-related (i.e. earthquakes).

The Social Pillar (S) measures a company's capacity to create trust and loyalty with its workforce, customers, and society by means of the best management practices. The Social dimension explains how private and public entities are affected by and manage health and safety and how they use, conserve, and develop their human and social capital to ensure their social license to operate or mandate to govern. The social perspective expresses the company's reputation and the health of its license to operate, which in turn are crucial factors in determining its ability to create shareholder value in the long term. The social considerations refer to issues of inequality, health, inclusiveness, labor relations, investment in human capital and communities, as well as human rights issues.[18] Hence, social factors that are related to the rights, well-being, and interests of people and communities and that may have an impact on the activities of

---

[18] See EBA, EBA Report on management and supervision of ESG risks for credit institutions and investment firms, EBA/REP/2021/18, § 2.4 Social factors and social risks, 75.

the companies' counterparties are increasingly being considered in the business strategies and the operating frameworks of businesses. Regarding these social issues, banks' interactions with the various stakeholders imply a structural exposure to numerous risk factors. We identify the following social credit factors:

- Health and safety includes risks and opportunities incurred across the life cycle of a product or service, or those related to relevant health and safety regulations and voluntary codes; occupational health and safety, public health and safety issues, as well as pandemics, wars, and conflicts. Health and safety may be considered an element of either social capital (public health and safety) or human capital (employee health and safety).
- Social capital regards risks and opportunities related to how companies treat and respond to consumers, citizens, and communities through providing responsible, affordable products, and services, including data privacy and security. Social capital factors can involve responses to socio-economic and demographic issues.
- Human capital comprises considerations such as a company's approach to treating employees through employment practices (hiring, recruitment, remuneration, development, retention, and related practices), and the working conditions it provides, including labor relations and employee health and safety. Human capital management within the value chain may materially affect credit through reputational impacts.

The Governance Pillar (G) describes the organization and its impact of decision-making at all corporate levels. The governance themes refer to the corporate governance and managerial structures that ensure inclusion of social and environmental considerations in the operations. Governance considers the system of rules, procedures, statutory frameworks, and practices by which companies are organized, how they make decisions, comply with the law, and align interests of the company with stakeholders. This includes how the entity manages strategic risks and opportunities to navigate potential disruptions. Governance dimension has always been one of the main areas for interest by supervisory authorities and especially in the recent years it has become a crucial variable for the sound and prudent management of financial institutions as well as for the stability of the entire financial system.

The governance pillar covers governance practices and in particular the inclusion of ESG factors in policies and procedures, identifying the impact of governance factors through physical and transition risk channels. The governance factors can lead to governance risks in many different ways. For instance, a poor code of conduct or a lack of action on anti-money laundering in a company can typically prevent its financial and non-financial resources, thus influencing its potential to perform. In these circumstances, customers and investors can lose trust and faith in the company, eventually leading to penalties and legal fees and impacting on its ability to conduct business over the longer term. Within the governance perspective, we identify the following governance factors: governance structure, risk management, culture and oversight, transparency and reporting, and any other governance factors as follows:

- Governance structure refers to how a company is set up to make decisions to meet the interests of its stakeholders. The independence, composition, and effectiveness of decision-making bodies and the institutional setups of sovereign and local governments may be key aspects. This also may include turnover, skill sets, and key person risk interconnected with the management structure.
- Risk management, culture, and oversight refers to how an entity makes and executes decisions or policies. This factor may include a company's effectiveness in identifying, monitoring, selecting, and mitigating the corporate risks, and the extent to which these risks are comprised within the company's risk appetite. These factors may also be traced back to the alignment of the company's culture to strategic vision and the company's ability to set and achieve strategic objectives within its capabilities.
- Transparency and reporting regards the extent to which stakeholders could access to all relevant information about a company, comprising its audited financial reports. This may include the quality, reliability, frequency, and timing of the company's information and the related standards of disclosure in its jurisdiction.

CHAPTER 3

The ESG Framework in the Banking Sector

Abstract The banking sector is increasingly aware of the need to integrate ESG factors into corporate strategies and business. In this context, banks must assess their strategies concerning ESG issues and their impact on revenues and risk in the perspective of sustainable finance. This chapter focuses on how banks can incorporate ESG themes into decision-making. In particular, the analysis highlights the steps to take in developing and implementing ESG approaches for investment and credit strategies and client engagement. The general purpose of the chapter is to clarify ESG concepts from a strategic perspective and to enable the assimilation of the upcoming regulation in this regard into the bank's value creation model. A step-by-step ESG integration process involves identifying relevant ESG issues, integrating ESG into decision-making, setting targets and goals, assessing current ESG performance, engaging with stakeholders, and measuring and reporting ESG progresses. Since the stakeholder model represents an emerging framework for the strategic vision of a bank, ESG metrics can be used to assess and measure bank performance and its relative positioning within a broader set of bank stakeholders. Hence, the chapter illustrates the interconnections between ESG strategy, the stakeholder model, and the creation of bank value; in doing this we provide a framework for understanding what ESG means for key areas of value creation across a bank.

E. Menicucci, *ESG Integration in the Banking Sector*, https://doi.org/10.1007/978-3-031-81677-2_3

Keywords ESG framework • ESG strategy • European banking sector • ESG challenges

3.1   INTRODUCTION

Sustainability is gaining importance in society and considerable interest for issues such as climate change, social inequality, or corporate misconduct are changing rapidly the financial market environment. Investors worldwide are showing a significantly amplified demand for sustainable financial products and banks have long been concerned with sustainability. Banks have to adapt to the changed market requirements and integrate regulatory requirements regarding sustainability into their frameworks. Sustainability is expected to affect banks along their entire value chains both from strategic and operational perspectives and to create new opportunities. Thus, the trend toward sustainability has the potential to drastically transform the complex banking sector. Since sustainability is influencing the reputation and business success of financial institutions, banks have acquired an increasing awareness of the need to integrate ESG factors into corporate strategies. ESG banking is a concept used to describe the banking sector's increasing focus on environmental, social, and governance topics. In other words, ESG banking is a type of sustainable banking that considers ESG factors. The main goal of ESG banking is to support projects and businesses that have a positive impact on society and the environment. This can include investments in renewable energy, green economy, green infrastructure, affordable housing, and more. By encouraging both economic and social development, ESG banking can help to create a more inclusive and sustainable future for all. In this regard, ESG banking also aims to improve financial inclusion (i.e. maximizing the impact of inclusive finance for low-income and vulnerable individuals[1]) and promote social responsibility to achieving sustainability goals and to creating more inclusive, resilient, and green futures.

---

[1] Financial inclusion refers to efforts to make financial products and services accessible and affordable to all individuals and businesses, regardless of their personal net worth or company size. Financial inclusion endeavors to remove the barriers that exclude people from participating in the financial sector and using these services to improve their lives. It is also called inclusive finance because everyone can access financial services that can help them build wealth, including savings, credit, loans, equity, and insurance.

Based on all the challenges posed by upcoming considerations concerning ESG factors into business strategy and risk management, more can be done to set out practical initiatives to embed ESG issues into banking operating models. Waiting or doing nothing is not an option. Due to the puzzling overflow of information and assumptions about future regulatory changes on this topic, it is difficult for most institutions to develop a comprehensive strategy for ESG factors. Anyway, banks can no longer ignore ESG, and in this regard they are starting to integrate these factors into their business models and internal operational processes. Banks must assess their strategies in the context of ESG and its impact on revenues and risk, to transform themselves and reaffirm their purpose in society. The suggestions for financial institutions to proactively incorporate considerations of ESG risks are foregrounded as a chance to rethink strategy, plans, and internal control frameworks. This trend has been encouraged by the growth of sustainable investing, which has led to an increased demand for ESG-related products and services from institutional investors. Hence, acting early to integrate ESG risks into business can be used for active positioning in the banking sector. This chapter focuses on the initiatives that may facilitate the introduction and functioning of an ESG framework into the banking sector. Thus, responsible banking, responsible investments, ESG-linked financial products, and sustainable initiatives are individually discussed.

3.2   Acting on ESG Strategies

ESG issues as well as their associated opportunities and risks are becoming more and more relevant for financial institutions. These issues have proven to be hot topics within the banking industry as banks adjust their strategies and practices to have more positive ESG outcomes. In order to achieve a sustainable development, financial institutions are required to define and implement an ESG strategy. How ambitious a bank is on ESG issues will determine whether it develops an ESG strategy that complements its existing corporate strategy or a new ESG-compliant corporate strategy.[2] Doing so, the discussion of motivation behind incorporating ESG can represent

---

[2] See European Commission (2021), Development of tools and mechanisms for the integration of environmental, social, and governance (ESG) factors into the EU banking prudential framework and into banks' business strategies and investment policies, BlackRock / Financial Markets Advisory, May.

a starting point for the examination of upcoming challenges and necessary courses of action in the development and implementation of an ESG strategy. The incentive for considering the topic can range from purely economic, regulatory, and/or legal motivations to intrinsically driven social and/or ecological reasons. The grade of integration into the banking business model varies from little more than some corporate social responsibility (CSR) activities to embracing ESG as a core part of the business model. In this regard, ESG is not only about ethical principles but it is a massive and multidimensional challenge as well as an opportunity. The fast development of ESG awareness among authorities, regulators, governments, and the public identifies a phenomenon that can no longer be overlooked by financial institutions and particularly by banks. The time to act has come because the transition toward a more environmentally and socially sustainable approach has become a pressing business imperative for banks and other financial services institutions. In particular, the financial services sector plays a crucial role in making this transition actual through the mobilization of capital. In this regard, some banks have already undertaken an ESG path by actively promoting sustainable financial products and choosing to stop or drastically reduce funding for certain sectors. How could individual banks respond to such a call for action? What does this mean for the future of banking? For a successful ESG strategy, a bank needs to define a roadmap for including ESG matters into the bank's business model and for adapting the organization to meet these goals. As ESG becomes gradually primary, banks must develop a structured approach that envisages a balanced set of effective ESG initiatives personalized to their own specific objectives.

In this regard, a bank should consider the ESG dimensions in the business strategy, even adjusting the products and customer portfolio in this sense, and in the organizational setup/governance. Furthermore, a bank should offer sustainable financing to customers and consider ESG risk in pricing and risk management. This is possible through the identification and the classification of sustainable assets, the consideration of ESG risks within the capital charge, the incorporation of ESG criteria in the distribution process (including MIFID), the implementation of an ESG Data Management, the reporting of own ESG risks and their impact to supervising authorities and stakeholders.

In the context of banking activities, banks should firstly start to weigh the implications that ESG factors may have on their core operations and business model in order to map and prioritize the key functions to be treated (e.g. product governance, investment processes, risk controls, etc.)

in line with the market trends and publicly available regulatory guidelines. Secondly, banks should define a strategy and an ambition plan for their move toward an "ESG-compliant" organization. Banks may choose to adapt their business and operations to ESG purposes, including ESG factors into their investment processes. Lastly, banks should develop an action plan with effective measures and milestones to improve their ESG in order to pose themselves on the market as ESG innovators and draw new clients or investors.

Banks need to respond to these ESG challenges with the following actions that identify consecutive steps to advance ESG initiatives.

1. Top executives should state the grade of motivation for their ESG strategy and set short-, medium-, and long-term ESG targets. Corporate priorities could change as ESG is indorsed to the top of the strategic agenda of the bank. They should also recognize the existing strengths and weaknesses of the bank to better evaluate when and how ESG targets may be achieved (e.g. the revision of the business strategy in relation to their target customers, new product, etc....).

2. Top executives must be stimulated to further take and roll out ESG initiatives, supported by strong ownership, governance, and culture. They might consider what steps they can take to increase their ESG scores and then to reach a competitive advantage. In this regard, strategic opportunities of ESG actions should be assessed with respect to how simple they are to implement, how impactful they are, and how important they are for the most significant stakeholders. Executives should make a concrete identification of what their purpose is and how they are going to make it happen to reduce the gap between historical performance measurement (e.g. return on equity and return on investment) and the new longer-term goals regarding ESG. To guarantee ESG success, top executives should set targets, align choices to ESG objectives, and track and report these actions back to stakeholders to meet their needs while creating value. The implementation of the ESG framework along the entire value chain of a bank implies the execution of these changes and transformations rolling out strategic initiatives supported by a strong leadership.

3. An action plan with clear steps and a relationship with the final ESG goals should be developed. The implementation of ESG dimensions could be an element of differentiation or a source of cost, particu-

larly where it is not aligned with a bank's corporate strategy. While some banks may be able to satisfy stakeholders' needs and attain attractive returns, other banks may experience an increase in costs without the expected impact in terms of return. Some banks still may not turn on ESG, then avoiding the costs but witnessing the reduction of revenues and the increase of risks over time. The execution of ESG strategies (e.g. responsible banking, responsible investment, ESG-linked financial products) has an influencing role both in launching new ESG products (i.e. compliant savings products) and in financing clients that launch innovative projects (e.g. environmental start-ups). Hence, banks must first make decisions based on organizational purpose and values and then set an approach that reflects their ESG principles and commitment. Effectively, these initiatives must come from the top and then cascade down to the rest of the workforce through the middle management.

Banks will need to be pragmatic in rolling out and executing ESG initiatives, sharing promptly progress and outcomes with stakeholders. While some initiatives could be more short-term focused, often determined by regulatory timing, others can entail a multi-year phased approach. Anyway, the achievement of a successful implementation depends on the belief and the behavior of the bank top executives that can enable a purpose-orientated culture. In this context, when the coherence of the ESG framework with the corporate strategy is low, top executives with a little level of commitment (skeptical approach) simply comply with the minimum regulatory requirements, while those with a high level of confidence in ESG principles (strategic approach) primarily adapt their organizations to fully incorporate ESG actions into bank strategies. A skeptical approach rejects ESG as a strategic priority, including ESG minimum regulatory standards in bank strategy while the ESG-focused strategy considers the dimensions of ESG as the main drivers of the bank's purpose even if they are in contrast with the bank strategy. When the coherence of ESG issues with bank strategy is high, the pragmatic strategy implements minimal ESG targets aligning with the corporate strategy if the commitment to ESG is low. On the contrary, a tactician strategy integrates ESG perspectives aligning with the bank strategy when the commitment to ESG is high. To reinforce ESG-related practices, this designed multi-approach makes available a clear template to implement these banking strategies. Banks should map out their actions on a standard squared matrix to highlight ESG strategies to perform (Table 3.1).

Table 3.1 ESG framework for banks and strategic ESG opportunities assessment framework

Environmental Impact of a bank on the environment or the impact of the environment on a bank	Products and services	• **Lending** to support development of renewable energy assets, (e.g. solar and wind farms), reducing emissions and exposure to high carbon assets over time • **Origination and distribution** of ESG-compliant bonds (originate green financial products, e.g. green bonds) • **Project financing** compliant with the Equator Principles (EPs)[a] • **Sustainable-linked loans**, e.g. loans for house enhancement to rise energy efficiency

	Risk management and regulation	• **Identification and quantification of climate risks** through heat mapping, scenario analysis, and stress testing
		• **Integration of environmental risks and opportunities** through a risk framework:
		  – Include climate impacts in corporate risk models and pricing
		  – Develop and implement an ESG risk management framework
		• **Regulatory requirements' scanning** to preserve compliance in all relevant financial markets

	Operations and supply chain	• **Carbon emissions** from offices and data centers • **Waste to landfill and incineration,** e.g. non-recycled computers, servers, photocopiers, and general office waste • **Business travel** including emissions from flight and train travel and vehicle fleets • **Non-renewable resource** used for electronics and office materials (e.g. materials used in computer and server production, paper and office consumables)

(*continued*)

Table 3.1 (continued)

Social
Contributions
of a bank to
fairness in
society
	Workforce	• **Diversity and inclusion** across employee
		categories (e.g. gender, ethnicity), social mobility,
		and pay equality:
		– Eliminate gender pay gap across company in 5
		  years.
		– Set targets to improve workforce diversity in
		  respect of gender, ethnicity, etc.
		• **Health and safety** and **well-being** programs (e.g.
		  employee mental health support)
		• **Human capital development**, by:
		– Upskilling of employees and training provided
		  (e.g. digital analytics training for experienced
		  employees)
		– Providing digital analytics upskilling training for
		  employees

	Products and
	services
		• Considering **social impacts of borrowers** (e.g.
		  human rights)
		• Developing **tailored customer service proposals**
		  (e.g. inclusive banking products, **inexpensive**
		  financing **housing**, and mortgages for the
		  community
		• Promoting **responsible behaviors around
		  money**

	Supply chain
	and
	distribution
		• **Diverse and inclusive** supply base, through office
		  supply and professional services contracts (e.g.
		  catering and events)

Governance
Quality of
processes for
decision-
making,
reporting, and
ethical
behaviors
	Transparency	• **Providing accurate and timely reporting to
		  stakeholders** on corporate purposes, strategies,
		  financial performance, and ESG tax benefits (e.g.
		  implications from investments in ESG initiatives)

	Accountability	• Ensuring that top executives are **responsible for
		  performance** and risk management, across both
		  ESG and other decisions, and that pay is aligned
		  to ESG-outcomes within the bank
		  Voluntarily commit to higher-standard ESG
		  reporting

	Independence	• **Ensuring a proper independent oversight
		  through** board composition, board diversity,
		  gender parity in board membership,
		  remuneration, and limiting controlling
		  shareholders and concentrated voting rights

	Ethical
	behavior
		• Undertaking business in an ethical manner,
		  avoiding bribery and corruption
		  Implementing third party controls, databases, and
		  monitoring to decrease potential bribery

ᵃThe Equator Principles (EP) are a set of voluntary guidelines adopted by financial institutions to ensure that large-scale development or construction projects appropriately consider the associated potential impacts on the natural environment and the affected communities. EPs are intended to be a common baseline and risk management framework for financial institutions to identify, assess, and manage environmental and social risks when financing projects

3.3   ESG FRAMEWORK: AN OPPORTUNITY OF VALUE CREATION

The stakeholder model represents an emerging framework for the strategic vision of a bank. In this regard, ESG metrics can be used to assess bank performance and the positioning on a range of relevant banking topics within a broader set of bank stakeholders as well as the financial metrics can be applied to measure bank performance for shareholders. This paragraph outlines, at a "conceptual" level, key questions and guidelines to ascertain a bank's readiness for—and the potential approach to—implementing ESG-focused metrics and goals in the context of the dynamic stakeholder model. Banks are no longer accountable just to shareholders but to a large set of stakeholders: for example, customers, employees, suppliers, communities, the press, public authorities, and regulators. These wider groups of stakeholders are increasingly involved in how ESG issues influence bank performance. From the ESG perspective, the stakeholders' environment includes internal and external stakeholders.

Internal stakeholders are:

- Management: top executives are required to integrate ESG into decision-making process and firm strategy and the executive remuneration is progressively related to ESG outcomes.
- Employees: since employee satisfaction impacts on productivity and it is positively associated with shareholders' returns, ESG can be considered an important control variable for talent management when employee attitudes toward their own work change.

External stakeholders are:

- Customers: growing demand for sustainable and green financing products (e.g. "green" mortgages), including ESG bonds, from corporate banking clients.
- Shareholders/investors: stakeholders and institutional investors are pressing banks to act on climate change and to include ESG considerations in investment decisions.
- Government bodies/regulators: the governments expect banks to embed climate risk management, to conduct climate stress tests, and to make climate-related disclosures mandatory.

- Communities and press: banks are gradually held to higher standards owing to pressure from lobbying groups and the press.
- Value chain partners: mortgage brokers and other intermediary partners are gradually interested in ESG strategies and banks that do not return to ESG effectually reduce partnership opportunities across the value chain.[3]

How do ESG metrics and strategies align with a stakeholder model? To answer this question, we must illustrate the connection of ESG strategy, a stakeholder model, and the creation of bank value. All banks need to balance the long-term interests of their stakeholders (including shareholders), making long-term investments and managing both short- and long-term performance trade-offs efficiently.[4] We believe that it is very important that a bank sets its ESG goals with the same relevance of financial metrics. Realizing ESG metrics in the bank-specific design process have to ensure that the attainment of the ESG goals will improve stakeholder value and not simply represent a "greenwashing" or "window dressing". For example, some banks may select to implement qualitative ESG incentive goals in addition to ESG factor data and reporting. Do ESG metrics and goals contribute to the bank's value-creation? ESG incentive metrics should be considered as any other incentive metric since they should support and reinforce strategy. Banks including ESG incentive metrics should plan an ESG framework allied with the corporate social responsibility and environmental strategies, reporting, and goals. Another essential factor in determining willingness is the measurability/quantification of the specific ESG issues. The case study for using ESG incentive metrics is to provide the management team with the tools to drive the implementation of initiatives that create significant segregated value for the bank or align with existing or emerging stakeholders' expectations. Banks must first confirm which metrics or initiatives could most benefit the bank's business and stakeholders. They must also identify exciting goals for these ESG metrics to growth the likelihood of overall value

---

[3] About the plurality of actors interested in banking activities (employees, customers, local communities, etc.), see Freeman, R.E. (1984), Strategic Management: A Stakeholder Approach; Pitman: Boston, MA, USA, pp. 35–42.

[4] See Kay, I.T., Martin, B. (2019), "Are Share Buybacks a Symptom of Managerial Short-Termism? New Insights on Executive Pay, Share Buybacks, and Other Corporate Investments", *Pay Governance*, May 14; Lizanne, T. (2019), "Stop Panicking About Corporate Short-Termism", *Harvard Business Review*. June 28.

creation.[5] Since the banks fall across a variety of drivers and variables that affect the different ESG factors that are relevant to short- and long-term bank business value, there is no one-size-fits-all approach to ESG metrics.

Most banks have faced, or will need to face, how to incorporate stakeholders' ESG considerations in their operating strategy.

Banks have a wide spectrum of arrangements for the disclosure and the adoption of the ESG incentive metrics and goals. In this regard, we can categorize three typologies of banks according to their readiness:

- Banks that are prepared to set quantitative ESG objectives since they implement robust sustainability strategies, including measurable metrics and goals (e.g. carbon reduction goals, net zero carbon emissions commitments, employee and environmental safety metrics, customer satisfaction, diversity and inclusion metrics, etc.).
- Banks that are prepared to set qualitative goals because ESG matters have been recognized as significant factors to customers, employees, or other; these banks likely already have plans or goals concerning ESG factors and they have developing formalized tracking and reporting (e.g. LEED,[6] certified office space, diversity and inclusion initiatives, renewable power and emissions goals, etc.).
- Banks that are developing overall ESG strategies at an initial phase for environmental and social issues before considering to correlate incentive pay to these sustainable priorities.

---

[5] See Jeucken, M. Sustainable Finance and Banking: The Financial Sector and the Future of the Planet; Earthscan Publications: London, UK, 2010; pp. 45–72; Jo, H., Kim, H.; Park, K. (2015), "Corporate Environmental Responsibility and Firm Performance in the Financial Services Sector", *Journal of Business Ethics*, Vol. 131, pp. 257–284; Finger, M., Gavious, I., Manos, R. (2018), "Environmental risk management and financial performance in the banking industry: A cross-country comparison", *Journal of International Financial Markets, Institutions and Money*, Vol. 52, pp. 240–261; Laguir, I., Marais, M., El Baz, J., Stekelorum, R. (2018), "Reversing the business rationale for environmental commitment in banking: Does financial performance lead to higher environmental performance?", *Management Decision*, Vol. 56, No. 1, pp. 358–375.

[6] LEED is the acronym of "Leadership in Energy and Environmental Design". LEED is the most widely used green building rating system all over the world. LEED certification identifies a framework for healthy, highly efficient, and cost-saving green buildings, which offer environmental, social, and governance benefits. LEED certification is a globally recognized sign of sustainability achievement, and it is backed by an entire industry of committed organizations and individuals involved in a market transformation.

For those banks moving to implement ESG incentive goals for the first time, the following proposals can help to transform the idea of an ESG incentive metric in a concrete goal with the potential to create value for the bank, in line with any proposal process for applying incentive metrics:

1. Select goals aligned with value creation. Banks should usually apply a principles-based approach to appraise the most proper ESG metrics for the bank as a whole, assessing for example, the effects on the organization, processes, business operability, etc.). In this regard, the relationship between ESG metrics and the bank value creation is influenced by the sector, the business model, the insights and the preferences of customers and employees, etc.

2. Operationalize ESG metrics within the long-term plans. Banks have to consider how to better incentivize the achievement of strategic ESG goals, considering that generally sustainable goals suit best into the framework of a long-term incentive.

3. Identify quantitative goals against qualitative milestones. The availability and the quality of data from sustainability or social responsibility reports usually establish whether a bank can set well-defined quantitative goals by setting annual milestones personalized to designated goals.

4. Determine employee involvement. Generally, ESG-focused metrics would be fitted for officer/executive level roles, since the group of top managers sets corporate policies that affect the achievement of quantitative ESG goals. Alternatively, banks may decide to project specific ESG incentive metrics to support the positive benefits of the bank's ESG strategy on employee engagement or to design a whole-team approach to achieve goals.

5. Consider the willingness to implement ESG goals in annual and long-term incentive plans. The most significant ESG incentive goal should be applied as biased metrics in balanced scorecard annual incentive plans. However, we wonder on whether some goals (mainly the greenhouse gas emission goals) may be better realized in long-term incentives. There is no right answer to this question because some quantitative goals are best set on an annual basis in line with the bank developments; other banks may have a strong long-term plan for which longer-term incentives are a better fit. These considerations and the interest in the value creation process have nurtured a debate on whether and how to comprise ESG metrics in incentive plans.

Below, we illustrate some key guidelines for applying ESG metrics in executive compensation incentive programs. In line with the continually evolving regulatory framework, the adoption of an ESG approach promotes the creation of value by causing ESG metrics leading to potentially significant benefits, if properly addressed, for banks. In particular, regarding the relevant challenges for banks, the main triggers of value creation can be recognized into the following considerations:

1. Regulatory optimization, which suggests arrangements to assurance compliance to the regulatory framework and the related ESG purposes with a view to creating value for all the stakeholders involved according to the operation of the banking activity.
2. Optimization of the ESG strategy standing at market level, in order to guarantee effective external communication of ongoing and future interventions with a sight on ESG transition. Additionally, a robust representation of the ESG risks could lead to reputational advantages and could rise the bank appealing for the main stakeholders.
3. Interactions with financial entities operating in emerging trends, considered as possible actuators of the ESG transition (e.g. entities operating within the Fintech industry).

Along with the considerations listed before, we have identified the following three key corporate areas for value creation in banking activity:

- Bank strategies. Integration of ESG factors within the assessment of business strategy and the identification, in the short term, medium, and long term, of the ESG risks to be incorporated in business and management strategies. For this purpose, the integration of the ESG data collection process through the identification of the information is necessary. This dataset must include information and data to support the preliminary investigation phase, with particular attention to the evaluation process of clients, as well as to the subsequent pricing, decision-making, management, and monitoring post-dispensing phases. In order to integrate the existing dataset investigated on a preliminary basis, the preparation of a questionnaire quali-/quantitative that the bank can submit to the customers at this stage is crucial to assess the sustainability characteristics of the operations underlying the financing and to punctually attain all necessary information for the subsequent evaluation and pricing steps of the operation.

- Management and monitoring risk process. In addition to the reflections regarding business strategies, the ESG factors and risks must also be integrated into the scope of the process of identification, evaluation, and monitoring of the risk. The relevance of such involvement stands up from the potential impacts of ESG risks on financial risks. Regarding the potential ESG adaptations of the lending process, in order to integrate and complement the developing activities from an ESG perspective, banks must revise the lending process through dedicated interventions in each phase of it in order to continuously manage the ESG aspects in the business process and punctually acquire all the sustainability assessments of the necessary information.
- Evaluation process of the counterparty. Review of the process, identifying the selection criteria related to ESG characteristics (e.g. business model and corporate governance oriented toward sustainability and ESG themes, share of investments in "green" projects, etc.), as well as evaluating the possible introduction of an ESG scoring or rating acquired from an external provider or processed internally. Finally, integrate current drivers' and KPIs' evaluations to consider the ESG components.

3.4   The ESG Challenge for Banks

In this paragraph we take a look at ESG's impact on banking activities. The ESG challenge is a huge change that requires a new approach and a broad transformation across the entire organization of a bank. Integrating ESG factors into key business banking processes supports the transition of the financial sector toward a more sustainable economy, considering that sustainability is becoming a core factor for banking operations. This trend also reflects bank stakeholders' increasing awareness on the importance for banks to incorporate ESG in their corporate strategy and a greater interest of customers, employees, regulators, and investors for sustainability. Hence, ESG appears in the foreground of the agenda because almost all banks have implemented firm-specific ESG strategies, which combine quantitative and qualitative targets (e.g. targets to reduce greenhouse gas or CO_2 emissions, targets for renewable electricity sources and paper consumption, bans on financing certain businesses such as coal energy, etc.). The ESG strategies differ among banks, because they can perform very detailed ESG strategies that consider ESG risk factors in the product development as well as in the pricing and sales processes. When a bank develops an ESG strategy, it is important to consider what ESG issues are involved throughout different dimensions as the following:

- The ESG impact on loans, investments, and other financial products. Set up new policies for investments that track the ESG concerns.[7]
- Bank revenues by developing new ESG products and services that improve brand loyalty, new revenue streams, and sustainable outcomes.
- Bank risk by assessing how climate and social factors may influence all the traditional risks (i.e. financial risk, credit risk, market risk, liquidity risk, reputational risk).
- Bank reporting, by providing disclosure and reporting for stakeholders, comprising risk modelling.
- Bank culture, by integrating an ESG approach in business processes and corporate governance to create a baseline mindset.

A bank should ensure that its approach to ESG is rooted by incorporating ESG issues into its banking operations and creating excellent and innovative ESG solutions for its clients. In this regard, the establishment of a unique coordination unit for ESG risks can be favorable but enhancing roles and responsibilities of existing units is even more beneficial. The ESG challenges impact differently on every area of the bank and the effective magnitude of the effects consists in planning a framework to fully integrate each ESG dimension in each business unit (i.e. corporate banking, investment banking, savings, mortgages, wealth, asset management, capital markets), designating a team for approaching through what ESG entails within the bank. ESG risks can impact all departments, divisions, and functions of a bank and in particular on the three different lines of defense model, that is, internal controls, compliance, and internal audit. The first line of defense impacted by ESG risks includes credit and trading business divisions, business and process owners, and operational management. The second line of defense comprises, but is not limited to,

---

[7] For example, ESG impacts may come from green bonds and social impact bonds in capital markets, from products that allow customers to align investments with their ESG interests in wealth management; from lower interest rate mortgages for energy-efficient homes, carbon-impact calculators or carbon offset services in retail. When it comes to ESG impact, in corporate banking, loans should be correlated to the achievement of social or environmental results.

compliance, risk monitoring, and business continuity management (BCM) functions. Starting from a correct risk inventory, risk monitoring must identify the processes and methods for addressing ESG risks and then the results must be included in risk reporting. Compliance in turn has to examine if the bank meets mandatory or voluntarily introduced ESG guidelines. After all, BCM must regard ESG risks as a trigger for business disruptions and provide for continuity. Internal audit as the third line of defense has to make sure that all relevant processes consider issues of ESG risks adequately and that they are treated consistently.

In general, dealing with ESG risks has become an embedded activity in all relevant bank processes since climate change and environmental and social factors are able to bring tremendous risk to a bank's assets and reputation.[8] This consideration is especially based on the impact of ESG risk factors on financial and reputational risks. For instance, clear decision criteria and control mechanisms must be anchored in the lending process: ESG factors have to be checked and assessed during the lending, similar to the evaluation of reputational risks in the KYC (Know Your Customer) process. This means that the assessments must be implemented initially when bank granting loans and then they must be carried out regularly and systematically by surveying all corporate customers. For example, the inclusion of ESG factors and ESG risks into the lending business affects the entire lending process (from risk governance to credit analysis, credit origination, credit monitoring, and ongoing client engagement). In this regard, the EBA has issued guidelines on loan origination and monitoring,[9] where it clearly specifies that ESG factors and correlated risks should be incorporated into the lending processes. Hence, ESG factors and risks can be combined into each step of the lending process and as the EBA recommends, "institutions should adopt a holistic approach".

Regarding risk governance, the real challenge lies with the tone that should be given from the top management to ensure that the credit risk culture meets ESG factors and to guarantee that the credit risk appetite incorporates these factors. A broad approach can be assumed for credit risk analysis to combine ESG factors into the credit assessment by

---

[8] See Konar, S.; Cohen, M.A. (2001), "Does the market value environmental performance?", *The Review of Economics and Statistics*, Vol. 83, No. 2, pp. 281–289; Forcadell, F.J., Aracil, E. (2017), "European Banks' reputation for corporate social responsibility", *Corporate Social Responsibility and Environmental Management*, Vol. 24, No. 1, pp. 1–14.

[9] See EBA (2020), Guidelines on loan origination and monitoring, May, EBA/ GL/2020/06.

assessing the borrower's exposure to ESG factors, business model, activities, and sustainable strategies. In this regard, the banks' ability to successfully classify, evaluate, and monitor the ESG business profile of their borrowers is the baseline for setting up-to-date ESG pledges and tracking the operating compliance with them. Bank clients or potential clients often have to undergo an ESG assessment process or a due diligence; in particular, these circumstances respectively happen when either renewing credit to an existing client or extending credit to a new client. ESG factors can be directly included into the client onboarding process, since the bank assesses the counterparty's profile and any disagreements. Specifically, the due diligence is a preventive approach because it helps banks to avoid or if anything to face opposing impacts related to the environment and corruption risks, human and labor rights associated with the clients, as well as to prevent financial and reputational risks. The due diligence approach leads to the ESG categorization of the client, which ponders the ESG profile of the counterparty as the main factor. Other factors to be valuated are the client's vulnerability to physical and transition risks and the assessment of the client's future plans to contain ESG risks. Finally, a central part of the process is the client engagement, which is relevant for inciting sustainable practices and supporting the clients in their shift toward more sustainable business models, operating, technologies, and culture. This engagement policy balances the internal perspective (e.g. improving capacities and expertise) and the external perspective (e.g. connectivity between banks and stakeholders to reduce ESG risks), which are both crucial for ESG sustainable policies.

In addition to these considerations, ESG affects the following bank functions:

- Risk management: understand risks and factors related to ESG factors (e.g. climate events, transition risk and potential greenwashing, etc.).
- Finance: introduce carbon accounting and carbon budgeting, and track progress toward ESG commitments.
- Compliance and audit: meet new reporting and disclosure obligations.
- Human resources: renew focus on social priorities with more focused actions around diversity, accessibility, and inclusion; enhance the power and performance that comes from diverse perspectives and teams.

- Information technology and operations: support the business in delivering ESG initiatives through new data and systems.
- Disclosure: reporting of own ESG risks and their impact to supervising authorities and stakeholders.

With regard to reporting, strategic actions should bridge the gap between historical paths of measuring performance and the new ones characterized by longer-term concepts coming with ESG. A practical way to do this is the preparation of the integrated reporting. This uses the principles of the International Integrated Reporting Framework to integrate the financial and ESG information that is most material for business strategies and the value creation process. This reporting approach helps banks to communicate broader performance to stakeholders and to show the interaction between ESG metrics and bank financials. The same reporting implies the definition of a clear set of KPIs for business units and processes and an effective tracking of ESG data to measure success in this regard. Hence, banks must have a strategy for communicating the ESG goals to their stakeholders through the use of a sophisticated and transparent reporting system. In this way, ESG actions and associated data can become a competitive advantage in determining the right pathway to a successful transition if they are leveraged and reported correctly.

3.5   THE REAL EFFECT OF INTEGRATING ESG DIMENSIONS IN THE EUROPEAN BANKING SECTOR

Which are the risks and opportunities for the European banking sector and what will they be in the next years? Banks play a central role in promoting the reallocation of financial resources for supporting the transition to a more sustainable economic system. Implementing sustainability in the bank business and risk strategies as well as incorporating ESG risks into every step of the risk management cycle gives banks the potential to take advantage of the market opportunities. In this regard, the incorporation of ESG metrics into credit decisions has become a real and actual need characterized by a substantial complexity. The integration of ESG factors inevitably concerns the entire credit cycle, including planning and credit policies, commercial practices, disbursement processes, risk management systems, and recovery strategies.

Banks that are speedier at adequately integrating ESG factors into their investment processes, lending decisions and dialogue with clients will be able to reach a competitive advantage in taking the chance offered by the transition in terms of growth in high-quality lending, expansion of client services, and valuable management of ESG risk factors. On the contrary, the banks that will be slowest to adapt in implementing ESG definitely find themselves more exposed to ESG risks, and they will be penalized in their market positioning finding difficulties in governing the quality of their bank portfolios.

An appropriate incorporation of ESG risks in some bank processes has becoming crucial for improving profitability. In this regard, there are three processes in which banks will have to endure the greatest changes: credit processes, investment processes, and data collection and management. To manage these challenges banks need to consider the ESG issues specifically in product design, pricing, and sales decisions besides embedding ESG into risk frameworks. The transition to sustainability should foster a reinforcing of credit standards through the integration of ESG factors into the dataset used rather than lead to an overall decreasing of credit standards.

Regarding the credit processes, banks should be able to value firms with adequate creditworthiness that require funds to adapt their business models to sustainability and the ones that could have difficulties when starting the transition. Banks have to closely analyze the sustainability and credibility of transition projects by gathering ample information on the company's climate change objectives, assessing the coherence of financing projects, and periodically monitoring the appropriate allocation of funds. To finance the transition projects, the ability to identify the sectors of economic activity, counterparties, and projects most exposed to climate risks becomes crucial.

On the supply side, the gradual introduction of more disclosure requirements will drive banks to value financing operations likewise on the basis of sustainability. The financing of sustainable sectors and green projects will enable banks to upgrade their published ESG indicators which then may influence their ability to attract new financial resources. On the demand side, the major disclosure enforced by the CSRD on large and listed firms will increase funding provisions in sustainable activities characterized by lower credit risks for banks. Sustainability reporting (or ESG reporting) makes it possible to improve the transparency od companies' ESG information. In line with these circumstances, firms that are not subject to regulatory obligations will also gradually adapt their business models toward sustainability and to increase transparency voluntarily regarding their sustainability features to facilitate the provision of financial resources. In line with this perspective, in the asset management sector, the asset managers are modifying the investment processes and are fully re-designing their portfolios under management, in order to improve the amount of green compliant products.

How do European banks incorporate ESG criteria? Despite increased efforts by EU banks and supervisors, ESG integration continues to be at a preliminary phase; in this regard, the effective integration of ESG factors within the bank's risk management, business strategies, and investment policies needs to be accelerated. In recent years, banks have enhanced governance structures and established sustainability teams to drive a broad ESG integration. However, the inclusion of ESG dimensions within banks' risk management practices is actually very challenging because of the lack of a common definition of ESG risks across banks. Hence, the specificities for the assimilation of ESG risks into governance are not defined yet and differ across banks. Some banks already offer ESG-related products and services, such as green project finance and energy-efficient mortgages, but the integration of ESG factors across their business strategies, investment decisions, and their range of products and services (comprising off-balance sheet exposures) has yet to be completed.[10] In addition, quantitative indicators for measuring ESG risks are yet to be fully defined, with a focus on qualitative elements related to risk processes within banks. Along the same lines, many supervisors have issued guidance on ESG integration, covering areas like risk definition, governance, strategy, risk management, and disclosure, with common guidance seen as beneficial for harmonizing practices and increasing ESG risk awareness among supervised institutions.

Notwithstanding, the effective integration of ESG risks into prudential supervision still faces a long process forward among EU supervisors. EU Regulators are increasingly looking into ESG topics and they require banks to incorporate ESG into their risk management systems as well as to have robust governance arrangements, appropriate organizational structures, transparent and consistent lines of responsibility to identify, manage, monitor, and report on the ESG risks they are exposed to. In this regard, several EU Regulations also require banks to provide more ESG-related data reporting in the future. As known, the board members should generate long-term value by ensuring that the bank is aware of and able to operate in an ever-evolving risk context to meet the expectations of their stakeholders. To that end, the board members should guarantee that the ESG risk management system is consistent with the overall approach to risk of the bank and fully aligned with the business model and its value

---

[10] Portfolio analyses of ESG lending and investment activities are often confined to specific product types and sectors, primarily for high-risk sectors, with a focus on renewables and green bonds.

proposition to stakeholders. The ESG risk integration into risk models and stress testing is in the initial stage; many banks want to integrate ESG factors into their lending activities as part of a broader ESG strategy, but they cannot rely on an adequate monitoring approach to recognize ESG risks as material in the context of longer time horizons. With further development needed to refine quantitative approaches to consider the environmental and social impact of banking activities, debates exist regarding whether ESG risks should be considered as a separate risk type or as a driver of other existing risk categories. As a result, ESG pillars and correlated specific risks can be assessed separately or holistically. Within this new perspective, as of today the inclusion of a sustainability-related or green credit in bank portfolio is no longer to be evaluated exclusively in reputational terms for the bank, but it also takes a substantial value from a risk assessment, capital allocation, and supervisory reporting perspectives.

As a result, European banks are putting significant resources toward ESG objectives, considering that sustainability regulation, climate change, investor expectations, and consumer preferences have enhanced ESG risks and opportunities. This ESG imprinting regards (but is not limited to) sustainable finance strategy, internal controls and supervision, sustainable product development and labelling, ESG/performance linked compensation, climate stress testing, diversity and inclusion, cybersecurity and data privacy, reputational risk, and ESG assessment of clients and vendors. By exploring these challenges, banks also play a significant role in generating financial resilience to environmental risks and in supporting the economy's development toward the relevant environmental changes. For this reason, in the banking sector new loan policies are focused on transferred credit toward more sustainable sectors of the economy. By adopting different types of green banking practices, banks reduce environmental sustainability risks and mitigate their impact. For example, these practices refer to the allocation of credit to renewable and clean energy projects and to the mobilization of capital to green economy.

3.6   ESG Action Plan: Roadmap and Practical Insights

This paragraph aims to describe a target operating model, that is, a potential action plan for the ESG transition. In line with the main aspects a bank identifies and in the perspective of adjusting the operations to ESG targets, it is proper to trigger a specified project to cascade an ESG action plan by any individual bank, considering its specificity. We propose a potential design approach structured into three macro-phases which are expected to project and implement the action plan.

The first phase involves an assessment "as is". This phase consists in an examination of the ESG framework of the bank with respect to the identified ESG objectives and in comparison with the main market player (i.e. benchmark analysis). This valuation is essential to identify the main impacted areas and functions impacted in the bank and to recognize the gaps with respect to the regulatory framework. The second phase consists of the preparation of the action plan. Based on the previous assessment, the bank must subsequently identify some actions to prepare a specific action plan. The actions concern the following areas:

1. Strategic planning: design of a sustainability plan through the identification of initiatives to be integrated into objectives of corporate strategies for an ESG positioning.

2. Review of the business strategies and the credit process: re-definition of initiatives relating to product catalog, client segments starting and monitoring the process of credit, also in line with the related legislation (e.g. EBA Guidelines on origination and monitoring), as well as identification of possible actions to adapt the pricing process from an ESG perspective.

3. Revision of the Risk Management framework: identification of actions to adapt the risk-taking process with a particular attention to the definition of the RAF, ICAAP, ILAAP, as well as to a possible Recovery Plan.

4. Revision of the monitoring and reporting system and disclosure: identification of the activities to adapt the monitoring systems (e.g. dashboard integration), reporting and disclosure to the market (e.g. investment objectives in sustainable projects reported in the periodical reports and in the pre-contractual information).

5. Revision of the internal document set: based on the actions identified in the previous macro-activities, set mapping of related legislation and identification of documentation to be adapted. Furthermore, consistent with the adapted pricing process, provide a possible review of the contracts. In order to ground these identified initiatives in the various areas, it is appropriate, at the same time, to adjust IT systems and reference database.

The third phase involves starting and subsequent grounding of adaptation activities. The grounding of the initiatives must be consistent with the activities (and related timing) defined in the action plan. Regarding the role of an officer of long-term corporate performance, the board plays a critical role in ensuring that the bank's ESG strategy and culture are closely aligned with the business model, in considering ESG risks from various different perspectives in the banking risk management, in detailing the risk appetite by specifying the types and the degree of risk that the bank is willing to accept. Within the bank strategy, sustainability aspects are included either as a separate dedicated strategy or as a complementary strategy to current ones.

Following this action plan, material ESG issues are faced in the business model by assessing its resilience to sustainability risks. For that, the risk management function plays an important role by supporting the board in the identification and assessment of sustainability risks in order to realize a well-designed decision-making process and management of the risks. The board has the permanent responsibility for assuring that sustainability risks are sufficiently considered for the respective risk categories. The board guarantees that the purpose, mission, vision, values, and code of conduct comprehend ESG themes and set the "tone at the top" to foster ESG considerations in the culture. In this regard, the appropriate tone at the top contributes toward establishing the appropriate ESG risk culture. When assessing sustainability risks, the bank's risk appetite must also be included since it considers the types and thresholds of the risks the bank may take, or avoid, in order to achieve its strategic plan.[11] The board should define how the bank manages the risks that occur through its operations and relationships in order to ensure that risks are managed according to the desired risk profile. To understand and to evaluate the risks that arise from ESG factors, the board has to oversee internal stakeholders, for example, management's assessment of material ESG issues, risks, and opportunities to include in the bank's external reporting and disclosures, and external stakeholders through mandatory and voluntary disclosures to engage stakeholders in the process of ESG transition.

In addition to the involvements envisioned at the level of business strategy, product catalog and framework of risk management and monitoring,

---

[11] The borders for the definitions of meaningful risk appetite and tolerance are determined by risk capacity. Risk capacity is the higher amount of risk that a bank is able to support for achieving strategic and business objectives.

it is necessary to assess the potential revision of internal governance, which is part of the broader regulatory framework defined by the EBA regarding internal governance[12] and loan monitoring.[13] In the context of internal governance, the implementation of the ESG topics suggest that the first appropriate activity is the assignment of roles and skills to the members of the board and/or of the board committees in relation to ESG risks. To this end, the method of assigning responsibility can be differentiated according to a principle of proportionality, which permits to consider delegating ESG risks to a member of an existing committee, or prefer the set-up of a dedicated committee.

There could be three options for the nominating committee depending on whether ESG issues are discussed on a regular basis in board meetings and whether these issues are addressed systematically:

- Standalone ESG risk committee (at the board or management level)
- ESG themes within other committees
- Corporate Social Responsibility/sustainability teams

At the market level, it has been observed that the significant banks have oriented themselves toward the establishment of ad hoc Committees. Corporate governance principles recommend that banks consider setting up a specialized risk committee that can support the board in performing its obligations. For ESG risks this may be a desirable option, especially when they are salient and/or where the implementation and monitoring of KPIs around ESG imply a high degree of technical expertise, for example, ESG skills, expertise, knowledge and experience, devises appropriate educational programs, and incorporates ESG issues into the self-assessment process. However, the board needs to pay attention to how a bank's limited resources are allocated to address actual and potential risks in the most cost-effective way and how different board committees coordinate. Some banks have established committees that bring together internal and external experts (e.g. technical consultants on issues related to sustainability) to support the Board of directors in defining the ESG strategy as well as to advise opinions on complex operations with ESG implications.

---

[12] See EBA (2021), Progetto di orientamenti sulla governance interna, EBA/GL/2021/05, 2 july.
[13] See EBA (2020), Orientamenti in materia di concessione e monitoraggio dei prestiti, EBA/GL/2020/06, 29 may.

In the case there is an internal committee in the board of directors, banks may prefer to integrate the committee already existing with technical, internal or external, expert on sustainability issues. For this purpose, a key aspect to consider in the context of internal governance is the promotion and dissemination of ESG risk culture through two possible intervention methods:

- Propagation of the level of propensity to ESG risks, metrics, and limits to its related issues defined at all levels of the organization in order to implement and spread a culture of ESG risks to all personnel involved in hiring, management, and monitoring of these risks as well as in the workflow of granting credit through processes, dedicated internal procedures, and communications.
- Planning, development, and implementation of training programs aimed at improving staff awareness of responsibility regarding ESG risks. In that respect, the bank may choose to include ESG risks within already established committees. There are advantages to this option, since ESG risks are not assessed individually but as a cross-cutting risk that impact the well-known traditional banking risks. This option reinforces risk management integration without adding extreme complexity, but implies coordinated decisions on ESG risks across committees to avoid gaps or overlapping deliberations. The last option is the incorporation of ESG themes into the existing Corporate Social Responsibility (CSR) based on the existing experience and norms. However, the definition of CSR is still generic,[14] since it is a voluntary approach and a larger or closer definition is applied depending on the context. There is also the risk of minimizing the significance of the ESG incorporation in risk management due to the usually scarce expertise and knowledge on this topic within the CSR teams.

---

[14] See Liang, H., Renneboog L. (2020), "Corporate Social Responsibility and Sustainable Finance: A Review of the Literature", *European Corporate Governance Institute*, Finance Working Paper No. 701.

CHAPTER 4

ESG Risks into the Risk Management Framework

Abstract ESG risk management is becoming an essential component of banks' sustainability strategies. In line with the prudential approach to risk management, this chapter focuses on ESG risks that can thus be defined as the negative materialization of ESG factors through their counterparties or invested assets. ESG factors can have positive or negative impacts on banks through their core business activities. In the negative side, ESG factors may impact banks' financial operational, liquidity, and funding risks, in addition to impacts attributable to the bank exposure to its counterparties and invested assets. ESG factors lead to negative financial impacts through a variety of risk drivers. The chapter presents the definitions of ESG risks separately and describes the ESG risk drivers, their transmission channels, and how these can impact on traditional bank risk categories. The chapter deepens the importance of integrating ESG risks in the risk identification, assessment, and monitoring process, in line with the provisions of the competent European Authorities. The integration of ESG risks in banking activities can be divided into two macro-phases: assessment and evaluation of exposure to ESG risks (which involves an "as is" analysis and the classification of exposures based on ESG variables) and then definition of the adjustment to the Risk Management Framework following the identification and the assessment of ESG risks. The effective strategies of banks should include ESG dimensions into bank risk frameworks considering that ESG risk is not a fully stand-alone risk type. Based

© The Author(s), under exclusive license to Springer Nature Switzerland AG 2025
E. Menicucci, *ESG Integration in the Banking Sector*,
https://doi.org/10.1007/978-3-031-81677-2_4

on a holistic approach to banking risks within the risk management process, ESG risks influence financial and non-financial banking risks through complex cause–effect relationships across them.

Keywords ESG risks • Bank risks • Risk management process • Risk management framework

4.1   Introduction

Environmental, social, and governance (ESG) issues as well as their associated opportunities and risks are becoming increasingly relevant for financial institutions, bank regulators, investors, and stakeholders in recent years. For banks, sustainability is not just an ethical question, but it generates a new type of risk: ESG risk. Hence, ESG risk is a new source of risk in the banking sector and it covers a range of qualitative and quantitative factors related to the sustainability impact of banks' activities and investments.[1] This chapter inspects the ESG risks that banks face, and explains how they are captured in the bank credit analysis. The relevance of ESG risks for the banking sector is principally due to the evolving regulations, policy measures, market developments, and new social attitudes that have gradually changed banking sector activities from satisfactory to challenged. Policymakers, consumers, and investors expect banks to play a key role in developing a sustainable economy and rising a sustainable finance. This opens up new lending opportunities for banks, but also exposes them to amplified risk of stranded assets, regulatory penalties, capital constraints, and reputational damage as well as a potential rise of credit risk and a distortion of risk measurement.

Banks should approach ESG risks in a holistic fashion when incorporating them into their risk management frameworks. These risks are also more difficult to be measured as they are mainly focused on subjective elements and all quantitative indicators are still to be defined. For this reason, ESG risks are considered systemic and can impact the financial system as a whole. Institutions need to build their resilience to ESG risks across different time horizons, by taking a comprehensive and

[1] See Kalfaoglou, F. (2021), ESG risks: a new source of risks for the banking sector, in Bank of Greece Eurosystem, Economic Bulletin, No. 53, July, p. 83.

forward-looking view, as well as early and proactive actions, under supervisory control. This process includes the adjustment of business and risk strategies and corresponding risk appetite statements, making roles and responsibilities fully transparent throughout all the three lines of defense.

Hence, risk management methods and processes must be amended, considering the complex cause–effect relationships across different risk types. This involves risk measurement and assessment techniques in run-the-bank and in change-the-bank processes as well as in stress testing applications. ESG risk is not a fully stand-alone risk type since it exerts influence on financial and non-financial risks in a bank to varying degrees. Banks are exposed to ESG risks directly, as well as indirectly through their loans and investments in their balance sheets, given their role as lenders and investors. The complex articulation of ESG risks imposes banks to become more selective in their business. Although ESG risks are separate and are typically managed as such by banks, they can have a relevant impact on a bank's credit profile. For example, a bank that fails to manage any of the three individual ESG risk factors may experience reduced demand for its market issuances, as investors worldwide gradually give high importance to ESG issues. ESG risks typically affect banks' creditworthiness through the same channels as conventional risks. Environmental risks primarily affect banks indirectly through their investment and lending decisions. They may also influence bank capital in future, as regulators may embed environmental considerations into banks' prudential requirements. Similarly, an inadequate commercial strategy and socially unacceptable business practices can both result in a loss of customers, hitting profitability. Social risks can also arise indirectly, for example, when the credit quality of a borrower weakens because of social considerations.

4.2   Mapping ESG Risks

ESG themes are a conceptual grouping of topics falling under each of the "E", "S", and "G" pillars. These pillars may translate into risks which collectively are referred to as ESG risks. Hence, ESG risks include environmental risk, social risk, and governance risk that influence banking activity both in lending and in asset class allocation as well as in profit and loss and liquidity. Starting from the EBA's definition of ESG risks, we explain some considerations about their evaluation and management, according to a holistic approach to ESG risks within risk management. According to

EBA, "ESG factors are environmental, social or governance matters that may have a positive or negative impact on the financial performance or solvency of an entity, sovereign or individual".[2] In its guidelines, the supervisory authority of the significant banks of the 19 Eurozone countries has issued its own guide on climate and environmental risks in which it sets out the expectations with regard to the management of such risks and the related disclosure to investors. The authority explicitly requests that banks incorporate ESG risk factors in the definition of their credit risk appetite and in their credit risk management practices.

The analysis of ESG risks is considered by supervisors as the new frontier to achieve a sustainable business model and a risk management system that ensures banks' competence for ESG-related challenges. The ESG approach to risks must be fully taken into account in the definition of strategies and objectives by integrating ESG risks in governance structures and managing these risks as drivers of financial risks. The analysis of ESG risks requires that banks must map all business units and divisions on the basis of ESG risks' framework applying the inside-out and outside-in perspectives. In line with the requirement of the supervisory authorities, banks have to reassess the impact of ESG risks on financial activity and the ways in which these risks translate into traditional risks (e.g. credit, market, liquidity). Hence, we analyze the concept underlying the ESG risks and their transmission channels toward the traditional banking risks. These risks may have different and specific features, due to their main causes and effects. For this reason, banks must classify ESG risks by considering separately the three factors: Environmental (E), Social (S), and Governance (G). Hence, ESG risks are divided into three macro categories for their identification and materialization into the risk management framework of a bank. The ESG risks must be inspected in all the above-mentioned perspectives according to a holistic approach for incorporating them into the risk strategy and the risk management cycle of a bank.

The "E" is about the way through which a bank influences and interacts with natural capital posing challenges on how to reshape the society

---

[2] See EBA, EBA Report on management and supervision of risks for credit institutions and investment firms, EBA/REP/2021/18, § 35.

to be climate resilient.[3] Environmental risks (E) reference to the financial impact of climate change and how these greatest threats to our planet influence the wider transition of the banking activities to more sustainable business models. Environmental risk has a double dimension since banks and financial institutions are both impacted by and contribute to climate risks. According to the Financial Stability Board, environmental risks are divided into two macro categories of risks since there are two main channels through which environmental risks can arise, that is, physical risk and transition risk. Many stakeholders are interested in these two dimensions of environmental risks and want to understand banks' strategies in financing the transition to a zero-carbon economy.

Physical risks refer to the financials of climate change, including more frequent extreme weather events and gradual changes in climate. They specifically identify natural catastrophes and the economic losses caused by them. In the last decades, the number of some types of extreme weather events has globally raised. Such events have become more possible and/or more severe due to the effects of climate change. It is also known that further warming will intensify climate risks and consequently the negative effects of them. Physical risk arises from damages to physical assets, natural capital, and/or human lives leading to output losses, as a result of climate-induced weather events. Physical risks include losses stemming from changes in physical capital because natural disasters destroy infrastructure and divert resources toward reconstruction and replacement. These risks affect also human capital, through deterioration in health and living conditions. The hard conditions due to the physical risks may have consequences on future expectations with a reduction of investment, given the prevailing uncertainty about future demand and growth prospects.

The physical risks are classified as "acute" if it is caused by extreme weather events (e.g. heat waves, droughts, floods, etc.) or "chronic" if they are due to progressive changes/shifts in climate patterns, for example, rising sea levels, rising average temperatures, and ocean acidification (e.g. increased temperatures, water stress, sea level rise, loss of

---

[3] These risks have been defined and detailed by the European Central Bank (ECB) in the "Guide on climate-related and environmental risks. Supervisory expectations relating to risk management and disclosure", November 2020. The Guide reports the expectations of ECB to which the banks should align themselves to consider the climate and environmental risks in the context of the current prudential framework. As concerns environmental risks, which are caused by numerous factors, banks must face both their physical impact and the effects of transition that overall identify the so-called "green transition".

biodiversity). For example, physical risks can directly cause damage to materials, a decline in productivity, or indirectly events such as the interruption of production chains. If there is no action to reduce the effects of climate change, physical risks will continue to increase in the future and the consequences of them can affect mainly market and credit risks as well as determine permanent deterioration in ESG target achievement with lasting adverse effects on banking activities.

When looking at "E", investors also face the so-called transition risks. Transition risks are those caused from the shift toward a low-carbon economy that aims to reduce the rate of climate change. This risk evidences the wider economic adjustment toward a low-carbon economy with the presence of a range of other factors, for example, the emergence of disruptive technology. Transition risk identifies loss financial charge that a bank may incur, directly or indirectly, following to the transition to a low-cost economy carbon emissions. This risk can happen, for example, for a sudden adoption of climate and environmental policies, from digitalization or from changes in preferences and trust of the market. Transition risks rise up if the business model is permanently threatened by systemic changes and its own negative ESG impact (e.g. the effects of political measures to combat climate change and their impact on manufacturers of combustion engines).

There are three main drivers of transition risks: (1) climate-related transition policies, such as the introduction of carbon pricing; (2) technological changes, in particular those contributing to energy transition and affecting the relative pricing of energy sources; and (3) shifts in consumer and investor sentiment or market patterns that can increase reputation risk.

Table 4.1 shows some examples of environmental risks divided by physical risks and transition risks for illustrative and non-exhaustive purposes.

Banks are exposed to climate-related risks through both their own operational impacts and the activities of their borrowers, customers, or counterparties. According to the outside-in and inside-out approaches, banks that provide loans or trade the securities of companies with direct exposure to climate-related risks suffer and accumulate climate-related risks via their credit and equity operations. In this regard, the ECB Guide represents a shared document that shows how relevant are a disclosed analysis of such risks to assure that banks are managed in a sustainable way.[4]

---

[4] The European Central Bank (ECB) underlines that "institutions are expected to incorporate climate-related and environmental risks as drivers of established risk categories into their existing risk management framework". The ECB focuses mainly on the "E" pillar. In some cases, the analysis converges on the ESG framework but the priority remains on the environ-

Table 4.1 Examples of climate and environmental risks

Risk factors	Micro-scope	Scope	Events (non-exhaustive examples)
Physical risks	Climate	Extreme weather event	Drought, floods, etc.
	Environmental	Chronic weather conditions	Pollution, water stress, loos of biodiversity, scarcity of resources
Transition risks	Climate	Policies and regulation	Unexpected adoption of climate policies, potential reduction of demand for properties in certain areas due to climate reasons
		Technology	Investments by energy supplier customers to revise their product mix from a sustainable perspective
		Market confidence	Customer preference toward products with an adverse climate impact
		Reputational	Investments in products with adverse climate impact
	Environmental	Policies and regulation	Unexpected adoption of climate policies, potential reduction in demand for properties in certain areas (e.g. risk of flooding)
		Technology	Customers originally engaged in highly energy-intensive sectors make investments in decarbonization
		Market confidence	Customers abandoning carbon-intensive goods and services
		Reputational	Financing of companies that carry out polluting activities

As the climate-related risks are really relevant, the ECB has stated that banks conducted a self-assessment according to the supervisory expectations included in the guide and that they develop compliant action plans.

mental issues, in line with the relevant interest in the academic literature and in policy institutions where the topic of environment is much more advanced and deepened in comparison with the other pillars of the ESG framework. See "Banking Supervision, Guide on climate-related and environmental risks. Supervisory expectations relating to risk management and disclosure", November 2020, p. 4. In line with this approach, see other prudential supervisors, e.g., Australian Prudential Regulation Authority (2021), Consultation on draft Prudential Practice Guide on Climate Change Financial Risks, April, and Bank of England (2019), Enhancing banks' and insurers' approaches to managing the financial risks from climate change, Supervisory Statement no SS3/19, April.

Table 4.2 Impacts of environmental risks on banking risks

Involved risks	Physical risks		Transition risks	
	Climate	Environmental	Climate	Environmental
	Extreme weather event Chronic weather conditions	Scarcity of resources Chronic stress Pollution	Policies and regulation Technology Market confidence	Policies and regulation Technology Market confidence
Credit risk	Potential impact on the value of the probability of default (PD) and related loss given default (LGD) relating to exposures to clients operating in sectors or geographical areas vulnerable to physical risks following to lower valuations of collateral in securities portfolios		High compliance costs and lower profitability resulting from energy efficiency standards, potentially leading to a decline of the customer's economic situation	
Market risk	Potential changes in market expectations resulting from serious physical events. These circumstances can lead to a reassessment of the risk and to a greater volatility as well as to a possible difficulty in accessing the capital market		Unexpected change in the price of securities and derivatives when transition risks upsurge	
Operational risk	Potential interruption of operations due to material damage to properties and data processing centers following extreme weather events		Potential reputational and legal liability risks following increased sensitivity to climate issues	
Other risk types (liquidity, business model)	Potential delays in the collection of invoices in the event that customers finance the repair of damages as a priority		Possible influence on the economic sustainability of some activities, causing a strategic risk for certain business models in the absence of the necessary adaptation or diversification interventions	

Source: ECB (See "ECB Banking Supervision, Guide on climate-related and environmental risks. Supervisory expectations relating to risk management and disclosure", November 2020. § 3.2)

The self-assessment plan represents the first step toward more accurate supervision of climate risk among all typical financial risks.

Table 4.2 illustrates the main impacts of environmental risks on the various bank risk categories as underlined in the guide issued by the ECB.[5]

[5] See "ECB Banking Supervision, Guide on climate-related and environmental risks. Supervisory expectations relating to risk management and disclosure", November 2020.

In the short and medium-long term, the impacts may occur through channels such as lower profitability, value lower than the property, a lower return of the asset, higher compliance costs, costs higher than legal standards, etc. Hence, environmental risks can produce negative effects on financial risk, in particular, on credit risk (e.g. impacts on probability of default (PD), loss given default (LGD), exposure at default (EAD) variables), on market risk (e.g. impacts on the value of financial instruments issued), on operational risk (e.g. in terms of legal risk and reputational risk), on liquidity and funding risk (e.g. value of financial assets and reputational aspects), as well as on reputational risk.

As shown in Table 4.2, physical risk may negatively affect credit risk through the lower collateral evaluation in real estate portfolios in areas that are highly exposed to climate risks. These circumstances, in turn, may negatively influence the values of the PDs and LGDs of these portfolios. Likewise, extreme climate events might shift the market behavior causing a potential market crash. Severe climate events might also determine physical damages to the banks' premises and properties as well as disruptions in the provision of services. In this regard, due to the severe potential negative impact of these events on operational risks, the management of ESG risks, as well as the inclusion of ESG risks into monitoring activities, may prioritize environmental issues within the wide management and supervision of ESG risks in banking. In fact, the EBA recognizes that the qualitative and quantitative indicators, metrics, and procedures issued for risk evaluation may be more complex for environmental hazards than those available for social and governance concerns.

The social pillar risk (S) is about the way through which a bank influences its various stakeholder groups (employees, suppliers, contractors, consumers, and others). The "S" pillar also allows investors to understand which stakeholder groups are most affected by the banks in their lending and/or investing portfolios and to assess how much these stakeholders suffer the negative effects of the corporate practices. The "S risks" typically include the risk of violating the human rights of stakeholders, that is, the limitation of workers' freedom of association that should result in a direct negative impact on workers' rights, the diffusion of social inequality, health troubles, or the exploitation of human labor, discriminating based on gender or ethnicity when hiring or promoting employees; lack of controlling whether suppliers or contractors pay a living wage to their workers; manipulation of customer data in a non-transparent and non-secure way; investing in projects or sectors with strong health implications such as the tobacco industry.

The concept of social sustainability has progressively been considered as an independent sustainability factor rather than a stand-alone part of a wider sustainable development. However, there is still ambiguity on the role of "S" in corporate frameworks and its integration into decision-making. In this regard, it is possible to contemplate two different perspectives:

- Addressing general social issues
- Addressing stakeholder welfare

Both approaches are partially manifested in social responsibility policies where banks incorporate social and environmental issues in their business operations and in their interaction with their stakeholders on a voluntary basis. Being socially responsible means both satisfying legal expectations and overcoming compliance by investing more in human capital, natural capital, the environment, and the relations with stakeholders. In this sense, in various circumstances, the responsibility of banks for the "S" pillar is identified in Corporate Social Responsibility (CSR) policies.

The governance pillar risk (G) relates to two components: corporate governance and business integrity. The first can really be considered as the way used by a bank to manage itself through policies, processes, and controls to achieve compliance and secure transparency in its dealings. Business integrity, on the other hand, regards the way to avoid corruption and bribery as well as prevent engaging with politically exposed persons (PEP) who may raise a reputational risk for the bank. The governance challenge has been crucial in the financial industry for over twenty-five years because good governance is considered one of the factors that permits to accomplish higher returns. For example, governance risks are triggered by corruption or similar issues in the board of directors of the bank or they come from an inadequate management of environmental risks as well as from a non-compliance with corporate governance frameworks/codes. Typical governance issues in the strict sense such as the role of the board of directors, the board composition, the board quality, the internal control system, etc., as well as issues in the wider perspective such as money laundering, know-your-customer (KYC) policy, client onboarding, etc., are addressed according to existing regulation, business conduct, and compliance requirements. New requirements must upsurge for the inclusion of ESG factors into bank governance. Thus, the board's quality and business ethics are of crucial importance. Also the board diversity is an important factor because all of its sub-characteristics (such as education or functional background, gender, age, race, or nationality) affect the board's leadership role and in the case of sustainability, its leadership toward the presence of ESG factors in decision-making process. The board of directors has the responsibility to incorporate environmental and social factors and to change the behavior to incorporate ESG themes within the overall recommendations for good corporate governance. This ESG inclusion has a compliance perspective for meeting regulatory requirements and a strategic dimension for comprising the entire set of ESG requirements in order to change operating from myopic short- to a long-term approach and create a long-term sustainable business.

The incorporation of governance issues in ESG analysis can take three perspectives:

- Governance from banks' own operational and organizational perspective
- Governance structures set up to implement and supervise the environmental and social risk framework throughout the bank
- Governance of the counterparties banks lend to

Bank governance is a long-standing issue which has been analyzed from several angles and which is dealt with from both a compliance and an ethical perspective.

The impact of governance risk is typically greatest in our assessment of asset risk, as a poor risk governance framework can lead to severe deterioration in asset quality. Governance risk can also materially affect profitability via fines or regulatory sanctions because of governance breaches. Environmental and social risks have historically had a low impact on banks' credit profiles, but their influence can still be material. Environmental factors can increase risk in banks' investment portfolios, and therefore lead to a deterioration in asset risk. If they cause financial losses, or if banks forgo profitable businesses on environmental grounds, they can also weigh on profitability. Social risks are less likely to affect asset quality, but they can have a material impact on profitability, for example, through litigation, regulatory fines, or regulatory measures that constrain earnings. Social factors can also have a positive impact on profitability, as in the case of banks which grow thanks to financial inclusion initiatives.

4.3   THE IMPACT OF ESG RISKS ON FINANCIAL RISKS

ESG risks can be defined as the negative materialization of ESG factors through their counterparties or invested assets.[6] From the negative perspective, ESG factors may impact banks' operational, liquidity, and funding risks, which are mainly affected by a bank's exposure to its counterparties and invested assets. The materialization of ESG factors has consequences on banks' performance because ESG risks materialize through the traditional categories of bank risks (credit risk, market risk, operational risk and reputational risks, liquidity risk and funding risks, etc.). Although ESG factors can have positive or negative influence on banks through their core business operating, our analysis focuses specifically on the last one, in line with the prudential approach of the risk management. In this regard, how do various ESG factors contribute to different types of financial risks between customers, service providers, and the bank? As well, have the sustainability risks complex cause–effect relationships with the individual types of financial and non-financial risks? Most banks view ESG risks from the perspective of reputational or strategic risk and they generally consider them transversal rather than principal risk types. Anyway, ESG risks materialize in known risk types (Fig. 4.1). For instance, if a bank grants a loan to a client that is suffering under the transition risk and costs of a green economy, its difficulties will affect the bank's credit position and credit risk. Further, extreme weather conditions can manifest through credit defaults and changes in market behavior in impairments. ESG risks therefore can affect counterparty, market price, liquidity, and operational risks. However, the cause–effect mechanisms require a wide range of banking expertise along the process, for example, on reflecting climate scenarios into business effects throughout the value chain.

Banks can be exposed to ESG risks from two points of view. The first identifies the direct exposure that arises from its own operations. For example, a bank may be subjected to operational risk if a branch of the bank is located in a high-risk flood geographical area. Thus, the environmental pillar of the ESG framework is migrated into a traditional bank risk setting. The second perspective regards the indirect exposure that upsurges from lending and investment activities. For instance, a bank may finance a counterparty that does not respect the regulation of safety in the

---

[6] See EBA, EBA Report on management and supervision of ESG risks for credit institutions and investment firms, EBA/REP/2021/18, § 2.2, 39–40

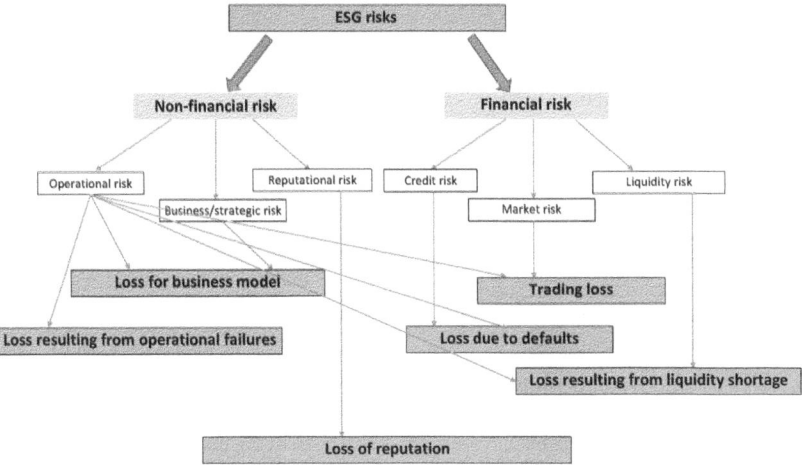

Fig. 4.1 Identification and materialization of ESG risks. (Source: KPMG; See KPMG, ESG risks in banks, KPMG International, 2021)

workplace and after an accident suffers for loss of customers and reputational risk, leading to an increased risk of default on the loan. Thus, the "S" pillar of the ESG framework is converted into credit risk. Further, a bank may invest in a counterparty's securities and a major fraud is found in the financial statements of that counterparty. Thus, the "G" pillar of the ESG framework is transformed into credit, market, and liquidity risks, reputational risk with consequences on bank profitability. Therefore, all ESG pillars should be turned into traditional bank risk categories. On the contrary, the climate risk may influence the value of financial assets causing losses for banks, investors, and financial institutions. In this circumstance, the losses are not directly caused by negative fluctuations of financial variables (i.e. interest rates or assets prices) but they are the effects of market risk. Hence, these losses are connected with the material disruption due to physical risk. With regard to credit risk, the effects are easier to be understood as they are the consequence of the difficulty or inability to refund loans, because of the physical destruction of assets or the death of human beings. It is evident that the picture of bank risks is large and complex due to the number and the composition of risks and their correlated aspects.

How can environmental, social, and governance risks impact financial risks through different transmission channels? In this regard, ESG factors

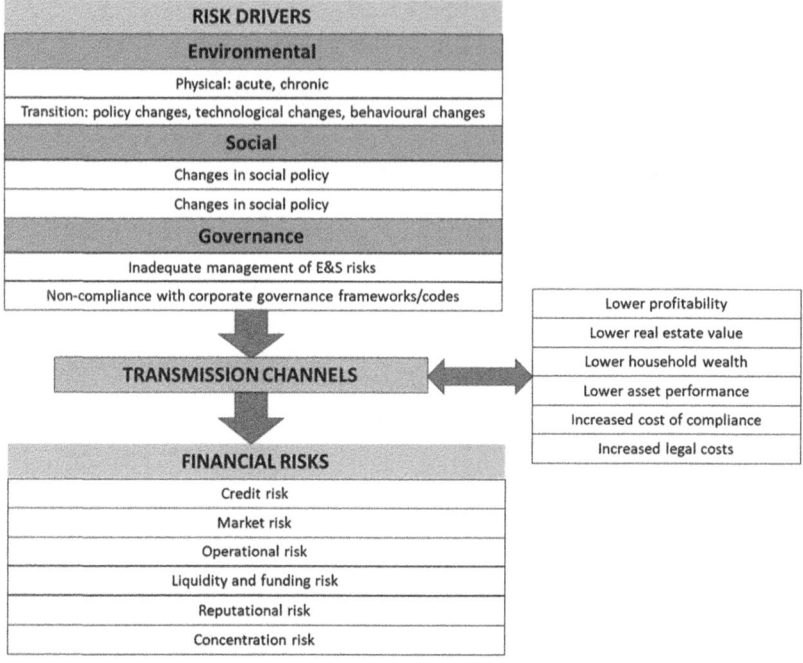

Fig. 4.2 Summary of ESG risk drivers, their transmission channels, and how these can impact financial risk categories. (Source: ECB See EBA, EBA Report on management and supervision of ESG risks for credit institutions and investment firms, EBA/REP/2021/18, § 2.2, 46)

generate negative financial impacts through a variety of risk drivers. Then, the causal chain that explicates how these risk drivers affect bank financial risks through banking counterparties and invested assets are named transmission channels. Figure 4.2 inspects the ESG pillars separately and presents the risk drivers and the transmission channels.

The causal chains that connect ESG risk drivers to financial risks are mapped despite their complexity. In other words, banks should explore the proper transmission channels that represent the way through which ESG risks might translate into a source of financial risk.[7] There are

---

[7] Basel Committee on Banking Supervisor (2021), Climate-related risk drivers and their transmission channels, Bank for International Settlements April.

microeconomic and macroeconomic channels to realize the concerned transmission. The microeconomic channels have a direct influence on banks, reflecting the effects of climate change on banks' operations or on the businesses, governments, and households they invest in or lend to. The macroeconomic factors have an indirect effect on the transmission consequences through macroeconomic variables, such as output demand and supply, inflation, economic growth, labor productivity, or market factors, such as commodity prices and interest rates. For example, a bank may face a growth in credit risk resulting (1) from an income effect, that is a reduction of borrowers' capacity to service debt, (2) from a wealth effect, that is loss causing by default on mortgage-backed loans when the value of collateral reduces, (3) from a transition effect, that is an increase in the probabilities of default (PD) and loss given default (LGD) of exposures within sectors or geographical areas that are subjected to the transition toward a low-carbon economy, or (4) from a sovereign effect, that is the effects from exposure to countries in which climate risk facts may affect tax and spending channels.

A significant process step in measuring and assessing ESG risks is the evaluation of the current ESG exposure. This includes the consideration of ESG risks while measuring capital adequacy as well as calculating regulatory and economic capital. The potential incorporation of ESG issues in banks' prudential frameworks mainly affect the assessment of the bank's capital adequacy and more generally of the bank's capital requirements. Also, Regulators may consider ESG risks into their supervisory work, for example, by comprising banks' resilience to climate change when conducting stress tests or encompassing reputational damage due to poor ESG performance. Similarly, the loss of clients because of ESG expectations disregarded may damage a bank's business volumes and therefore its earnings capacity, profitability, and liquidity. Then, clients may pull out their funds and the banks' issuances may attract less market interest whereas investors increasingly incorporate ESG considerations into their investment decisions. Table 4.3 shows the rating factors that design the bank's baseline of credit assessment, with a brief explanation of how the assessment captures ESG risks, directly or indirectly.

In addition to the negative impacts on institutions through their effects on counterparties, ESG risks can also influence the financial system and the economy as a whole, with potential systemic consequences. Negative impacts of ESG factors could affect macroeconomic factors, such as economic growth, government debt, gross domestic product, labor

Table 4.3   ESG risks related to rating factors

ESG	Rating factor	Relationship ESG-rating
Environment Social Governance	Solvency— Asset risk	Environmental factors influence banks' asset quality in terms of credit risk and market risk as well as the traditional financial factors. Bank exposure to environmental risk is generally low, but it can be material in cases of concentrated lending to individual sectors or projects. Risk governance determines banks' risk appetite, and therefore it is a key driver of asset risk. Poor risk governance can materialize in severe asset quality issues.
Environment	Solvency— Capital	Capital can be subjected to a variety of regulatory measures designed to capture environmental considerations (e.g. green supporting factor, brown penalizing factor). Bank regulators may also consider the banks' resilience to climate change risk in their stress testing methodologies.
Environment Social Governance	Solvency— Profitability	ESG factors can affect a bank's profitability. Governance inadequacy can expose banks to considerable financial penalties and banks with poor ESG metrics are exposed to the risk of losing customers. Social considerations can have a positive impact on profitability for banks which focus on financial inclusion to grow.
Environment Social Governance	Liquidity— Liquid resources and funding structures	Bank investors are increasingly incorporating ESG criteria into their investment decisions, pressing banks to demonstrate strong ESG credentials. This trend could spread to depositors, which may demand premium from banks with poor ESG credentials. The use of sustainable funding resources (e.g. green bonds) helps banks to increase their funding diversification.
Governance	Qualitative Aspects— Business model and corporate conduct	Although a separate score for corporate governance quality in bank scorecard is not assigned, governance considerations influence the score assigned to the scorecard factors. It is possible to adjust downward the financial profile of some banks due to corporate behavior considerations, where the risk function and governance framework are not adequate for the risk-taking, or when the compliance of the board and management oversight is poor. More exceptionally, a positive adjustment can be applied when a bank has a very strong approach to risk management.

productivity, and socio-economic changes.[8] This context, in turn, could have an impact on financial institutions by affecting the economy in which they operate and then their financial performance or solvency. Specifically, in relation to environmental risks, it has been advised that because of their scale, extent, and complexity, the impact of ESG risks could be systematic, thereby affecting overall credit risk and market risk. These risks could interact with each other, amplifying shocks and stresses, the latter of which could lead to spill overs that could simultaneously disrupt multiple parts of the financial system, which could in turn have an impact on the institutions' financial performance and solvency.

4.4   INTEGRATING ESG RISKS IN THE RISK MANAGEMENT PROCESS

This paragraph explores ESG factors and related issues in the banking sector, proposing a holistic approach to ESG risks within the risk management process. ESG factors and risks must be integrated into the process of risk identification, evaluation, and monitoring. Risk management is a key function within the banking activities and in this context all the traditional risk categories banks are used to dealing with (i.e. credit and counterparty risks, market risks, liquidity risks, operational risks, etc.) have specific effects on the bank itself. On the contrary, ESG risks are not fully standalone risk types that merely impact directly on bank functions, but they exert influence on financial and non-financial risks through composite and complex cause–effect relationships across different traditional risk types. Hence, risk management must consider new perspectives; in particular, the impact of ESG risks on the bank and vice versa the risks that the bank translates to its stakeholders and the environment due to its business activities.

Regarding Risk Management, ESG risks are significant, complex, and multiforme, ranging from physical risks such as climate change–related natural disasters to transition risks originated by legal and policy risks from greenhouse gas emissions as well as governance or social issues (e.g. human rights abuses). Hence, the incorporation of ESG risks into an existing risk

---

[8] See Engle, R., Brogi, M., Cucari, N., Lagasio, V. (2021), "Environmental, Social, Governance: Implications for businesses and effects for stakeholders", Corporate Social Responsibility and Environmental Management, Vol. 28, pp. 1423–1425.

management framework implies several challenges, for example, the long-time prospect, the high level of uncertainty, and the interconnections between lack of data and methodologies. Thus, a new area of investigation is needed with new tools and methodologies. By including ESG considerations into the risk management framework, banks can better predict and manage these risks, and then have a positive impact on bank financial performance. For example, banks that fail to adequately assess and manage climate-related risks could address stranded loans or lawsuits, which could then impact negatively on their bottom line. Regulatory Frameworks in Europe have taken note of these aspects and the EBA now requires banks to disclose multiple data-aspects regarding ESG risks in their risk reports (Pillar III).[9]

Following the traditional risk management path, as a first step banks should define the context and introduce the appropriate risk culture although this can be a difficult effort that contains norms, attitudes, and behaviors related to ESG risk awareness.

First of all, risks should be identified. Risk identification may begin with a risk inventory that can be appropriate for an initial risk screening and a starting point for prioritization. Thus, the risk identification starts by considering ESG risk factors in the risk inventory for exploring the risk landscape. Due to the wide range of interconnections between financial and non-financial risks, ESG risks cannot be assessed using a linear method. On the contrary, ESG risks have to be recognized by inspecting the cause–effect relationships and/or common triggers between them. ESG risks must also be analyzed for every risk type, that is, within each risk type an investigation must be made on the extent to which the specific risk is able to change the assessment of other risk type, preferably bearing in mind the second-round effects.

Due to the innovative characteristics of the ESG risks and the limited familiarity with them, a detailed description is crucial, focusing on the risks themselves, rather than discussing about an overall ESG issue (e.g. climate change). For example, an appropriate level of disaggregation, geographical location, the carbon emission intensity of exposures, the effect of alternative scenarios for transition and different time horizons should be inspected.

---

[9] See EBA (2022), Final draft implementing technical standards on prudential disclosures on ESG risks in accordance with Article 449°, CRREBA/ITS/2022/01, January 24.

The analysis can be enabled and developed if a root cause approach[10] is used in order to verify whether an additional step toward a more comprehensive risk assessment is needed. This approach consists of a further analysis of the underlying of the ESG drivers of business risk, their potential influence, and the potential remedial actions. Specifically, prioritization can help the bank to understand and address the urgency of the required reply, the types of necessary actions, as well as the amount of investments for the risk response. Regarding prioritization, this can be based on numerous criteria, including the ability of the bank to familiarize and respond to risks, the scope and the nature of the risk within the bank's portfolios, the rapidity and scale of risks' impacts on the bank, and finally also the institutional and financial ability of the bank to re-establish normality. This detailed analysis allows to separate the required changes, so that the bank can address a criticality at its source rather than face its symptoms. The results from this edited risk inventory process can be used as a basis for the construction of a consistent and detailed taxonomy. Based on these results, possible scenarios as part of capital planning and stress testing can be designed. This approach is not similar to the self-assessment of traditional risks since ESG risks are new and innovative with extensive impacts in breadth and magnitude to address depending on the own business, market, customers, risk profile, geographical location, and risk appetite of the bank. Given the relevance of ESG risk integration into the risk management framework, in accordance with the provisions of the European competent authorities, it is crucial to prepare a risk mapping and grounding the management actions dividing them into two macro-phases.

The first phase is the assessment of the exposure to the ESG risks in bank lending or investment portfolios. The assessing of the ESG risks is crucial for understanding and mitigating potential impacts on the bank's operations, reputation, capital allocation, and financial stability. Given that many of the risks are not typical and have complex and non-linear effects, the modelling of these risks is complex. Scenario analysis is particularly beneficial in this respect, allowing the exploration of a range of possible outcomes and the assessment of the evolution of the portfolio under

---

[10] The root cause approach is a systematic process for identifying the root causes of problems and then to respond to them. The root cause analysis is based on the concept that effective management should find a way to prevent problems before they occur and before they impact on the work of an entire organization. Tools for understanding root causes include the five whys, cause-and-effect charts, hypothesis testing, and comparative analysis.

different scenarios. In addition, this approach can familiarize the banks with the nature of ESG risks, increase their awareness and knowledge. For example, climate-related risk exercises are more forward-thinking, allowing banks to experience impact quantification of transition or physical risks in their portfolio. Stress testing for these risks can be very complex and less invasive than scenario analysis, but some indications may be extracted to assess the impact of climate risks on banks' business model and strategy.[11]

This phase starts by evaluating the bank's impact on ESG factors to identify areas of vulnerability and then it considers financial risks associated with climate change, such as physical risks, regulatory changes, and market shifts. These risks can have direct financial repercussions if not managed efficiently. Thus, by gathering and analyzing material information related to ESG risks, it is possible to develop robust risk management strategies and enhance the resilience in an evolving business landscape. In order to carry out an assessment of the ESG risks faced by the bank, the following methodologies can be summarized:[12]

(a) Portfolio alignment method, that is, the verification of the alignment of the "as-is" portfolio at different levels (e.g. customer cluster, sector, geographical area) with respect to the ESG targets set at an international level. The objective is the determination of the positioning of the bank with respect to international objectives and calibrate short-term interventions and medium-long term strategies.

(b) Risk framework method, that is, the evaluation of the current portfolio's ability to undertake ESG risks in the long term through stress and sensitivity analyses. The objective is the measurement of the impact of sustainability-related issues on the risk profile of the bank's portfolio.

---

[11] Given the potentially relevant challenges on financial stability, some central banks have performed climate-related stress testing exercises, extending the scope of these exercises also to new frontiers. For example, the Autorité de controôle prudential et de résolution (ACPR)-Banque de France exercise has a 30-year time horizon, considering a static balance sheet for the first five years and assuming a dynamic balance sheet onwards. It uses three scenarios: an orderly transition to 2050 goals, a sudden transition, and a late transition. The results are quantified mainly as credit risk manifestation and in particular as the increase in the probability of default of different sectors. See Banque de France, (2021), A first assessment of financial risks stemming from climate change: the main results of the 2020 climate pilot exercise, ACPR, No. 122-2021.

[12] About methodological approaches for assessing and evaluating ESG Risks, see EBA, EBA Report on management and supervision of ESG risks for credit institutions and investment firms, EBA/REP/2021/18, § 3.2.

(c) Exposure method, that is, the ESG assessment carried out directly at counterparty or exposure level with the purpose of complementing the standard assessment carried out on financial risk categories. This method is based on a direct evaluation of the performance of exposure in terms of its ESG attributes through a calibration at the specific bank level. The objective is to extract an indication regarding the degree of exposure of a portfolio to ESG risk, as well as the actions to do to make it more sustainable. This method can be used to complement the standard assessment of financial risk categories. Although this method is not based on a complex scenario analysis, it considers backward-looking metrics and makes banks able to classify their ESG risks' exposures to evidence the specific sensitivities of different segments and sub-segments of economic activity to ESG factors. This method suits well to all three aspects of ESG and it is considered the most appropriate if it is compared with the others. Within the exposure method, regulators have developed some methodologies that can realize the ESG risks measurement. We illustrate the following four methodologies:

i. ESG ratings. With this methodology ESG ratings are provided by specialized rating agencies. They are stand-alone ratings on ESG factors and they consider the risk exposure to ESG factors. Rating agencies measure also the capacity of the management to face risks and to take opportunities. These methodologies are generally based on a quantitative analysis of key issues identified for each company, but they also consider qualitative information collected by analysts from public reports and engagement with companies.

ii. ESG evaluations. The ESG evaluations are provided by credit rating agencies (e.g. S&P ESG) and these evaluations integrate ESG factors into the standard credit analysis. They measure how ESG factors affect both certain scorecard components such as cash flows and leverage, and/or variables outside of the scorecard. Although some difficulties arise in comparing ESG ratings issues by different providers since they apply different weights to the individual elements of ESG factors, these evaluations contribute to give additional input to the existing financial risk assessment.

iii. ESG evaluation models developed by banks in-house for their own assessment. These internal methodologies are generally developed by larger banks that organize their information systems on the basis of internal data deriving from wide data sets concerning their clients. These evaluations are internal and need the validation of regulatory authorities to be compliant with the existing rules.

The second phase consists of grounding the Risk Management Framework by adjusting, monitoring, and reporting activities following the identification and the assessment of ESG risks. The implementation of ESG risks into the risk management framework improve the risk management process with the addition of a new perspective that is the impact of risks on bank's stakeholders and, in turn, the effect of these risks on the bank's overall performance. Hence, banks should approach ESG risks in a structured manner supporting the establishment of a holistic ESG risk management that shed new light on the different ESG risk types and influence the correlated ratings.

In Table 4.4, we summarize some potentials actions in banking processes, to incorporate ESG risk factors within a Risk Management Framework.

ESG risks need to be embedded in all relevant processes. For instance, ESG criteria and control mechanisms must be contemplated into the lending process. In particular, ESG factors have to be checked and assessed during the lending process, similar to the examination of reputational risks in the KYC (Know Your Customer) process. This means that assessments must be implemented both initially when granting loans and regularly by surveying all corporate customers.

The internal control system,[13] identifying with the first line of defense role, typically already incorporates a broad range of non-financial risks but banks still will have to amend their risk management processes accordingly to the ESG metrics. This implies the enhancement of qualitative risk assessment methods and tools by including ESG risk-related aspects and

---

[13] The internal control system at a banking group level consists in a set of rules, procedures, and organizational structures with several goals. It ensures that the corporate strategy is implemented, it guarantees the achievement of the effective and efficient corporate processes, it protects the value of corporate assets, it ensures the reliability and the integrity of accounting and management data, and it ensures that operations comply with all the existing rules and regulations.

Table 4.4   Potential actions to adapt the Risk Management Framework

Risk processes	Internal governance	Application perimeter	Limits, metrics, and indicators
Risk Appetite Framework (RAF)	• Integration of the internal governance system by including the structures responsible for defining and monitoring processes and procedures related to the management of ESG risks.	• Adjustment of the perimeter relating to the bank's risk profile to integrate ESG risks.	• Potential review and/or integration to include risk-return objectives to be achieved (e.g. based on customer clusters) and the consequent operational limits (e.g. limits on counterparties that may not respect the ESG objectives, in line with the business). • Integration of the existing set of indicators to include key performance indicators (KPIs) and key risk indicators (KRIs) focused on ESG dimensions.
Internal capital adequacy assessment process (ICAAP)/Internal liquidity adequacy assessment process (ILAAP)		• Integration of ESG factors and risks within the ICAAP, ILAAP and Recovery Plan in line with the provisions of the RAF through the inclusion of a section dedicated to ESG risks.	• Qualitative and quantitative adjustment of ICAAP and ILAAP in relation to established thresholds and limits as well as stress tests. • Development of scenario and stress analyses over within a medium-long term time horizon (longer than typically expected timing, i.e. 3 years), to continuously evaluate resilience to ESG risks at portfolio, industry, and counterparty levels with different levels of aggregation depending on data availability.
Recovery Plan			• Integration of ESG risks within the basic and adverse scenarios, as well as of the set of indications adopted as part of the Recovery Plan.

concerns and by connecting them with non-financial risks (in particular, operational risk and reputational risk). The second line of defense includes, but is not limited to, risk monitoring, compliance, and Business Continuity Management (BCM) functions. Risk Controlling must develop methods, processes, and tools for dealing with ESG risks from the amended risk inventory and then for including results in risk reporting. The compliance function examines if the bank respects the legal or voluntarily ESG guidelines. Finally, BCM must encompass ESG risks as a trigger for business disruptions and guarantee for continuity. Internal Audit as the third line of defense has to make sure that all relevant processes contemplate ESG risks adequately and consistently.

The definition of internal processes and tools aimed at monitoring and reporting the exposures to ESG risks, with different levels of granularity (e.g. at the level of portfolio, counterparty, exposure), implies the identification of quantitative and qualitative indicators in relation to the ESG risks individually for the three ESG pillars within the evaluation and control framework risk. In Table 4.5 are reported potentials indicators (both qualitative and quantitative) for each ESG factor.[14]

Generally, banks have two paths for integrating ESG risks: the corporate governance structures and the risk management frameworks. Banks should adjust their management and oversight structures as well as internal processes, including the roles and responsibilities of the board of directors to contribute to the overall objectives of sustainable development by incorporating ESG risks and societal expectations in the corporate governance. The risk management path should embed ESG factors into existing frameworks and translate the ESG risks into accurate complete, simple, and comparable metrics, allowing the management to adjust business models and internal processes accordingly. The ability to effectively classify, measure, and monitor the ESG business profile of bank credit allocation, portfolio management, as well as investment advice to clients is essential to set informed ESG commitments and track progress against them.

A bank's corporate governance and risk management framework are also key drivers for the credit quality. In this regard, ESG risks have a long-term impact and if not properly addressed, they might negatively affect the

---

[14] Indicators designed according to the reporting principles and indices provided by the "GRI Standards" as well as in line with the Implementing Technical Standards on Supervisory Reporting (ITSs) published by the EBA regarding disclosure on ESG risks.

Table 4.5   Examples of qualitative-quantitative indicators

ESG factors	Quantitative indicators	Qualitative indicators
Environmental	Green Asset Ratio (GAR)[a] Greenhouse gas emissions and indirect emissions. Etc....	Evidence that the business strategy integrates the impact of climate and environmental factors and risks. Any awards/recognitions or certifications. Etc....
Social	Presence and share of activities that contribute to the mitigation of social risk compared to overall activities. Etc....	Presence of policies/procedures that highlight a direct or indirect commitment to reduce socially harmful activities. Etc....
Governance	Hiring rate by gender and position. Turnover by position, gender, and age. Composition of top management and the board of directors by gender and age. Etc....	Inclusion of ESG factors within internal governance processes and procedures, e.g. inclusiveness, management of conflicts of interest, internal communications. Presence of dedicated training programs. Etc....

[a]In the current context of growing attention to sustainability, banks are introducing new criteria to assess the creditworthiness of companies. Among these, the Green Asset Ratio (GAR) is one of the most relevant. The GAR is an indicator that was recently adopted by the European Banking Authority (EBA), following the approval of the European Taxonomy, a classification of economic activities considered sustainable in Europe, pursuant to the European Green Deal. The GAR measures the commitment of banks to sustainability and is calculated by comparing the nominal value of green financing to the total risk-weighted assets. Green financing is that intended for projects that contribute to the climate objectives of the Paris Agreement, such as the fight against climate change, energy efficiency, and renewable energy

solvency and liquidity position of a bank. As risk-taking institutions, banks need to ensure that their investment decisions are proportionate with their risk appetite by placing proper risk management structure, as well as strong compliance and control functions. Unlike environmental and social risks, which can be driven by external factors such as regulation or demographic change, governance risks are largely internal.

A sound governance structure is a crucial element for effective risk management processes and in this regard, a poor risk governance can lead to an asset quality deterioration, eventually resulting in bank failure. In this context also, ESG risks can affect all divisions and departments of a bank and the various areas of the three lines of defense model, including profit and cost centers. The establishment of a central coordination unit for ESG risks can be beneficial for banks as well as the enhancing of roles and responsibilities of existing units. For example, profit centers within the first line of defense can be affected by ESG risks and in particular the credit and trading business divisions have to consider ESG risk factors in banking product development as well as in pricing and sales processes. This consideration is especially significant with regard to the impact of ESG risk factors on financial risks and reputational risks.

The integration of ESG criteria into credit decisions is therefore a concrete and current necessity characterized by a considerable complexity. In fact, ESG dimensions necessarily concern the entire credit cycle, starting from planning and credit policies, and then involving commercial practices, risk management systems, and recovery strategies. Hence, ESG risks impact on bank ratings since they are captured in the bank credit analysis. As a consequence, ESG ratings are becoming increasingly significant when screening the investment landscape. Since ESG credit factors (e.g. investment in environmentally harmful industries, misconduct issues, and governance failings) can affect bank credit strength and can influence rating outlooks, they must be captured in the bank credit analysis. We define ESG credit factors as environmental, social, or governance factors that influence the capacity and willingness of an obligor to meet its financial commitments. The potential influence of ESG credit factors depends on how much they affect the capacity and willingness of an obligator in taking its financial commitments. This influence could materialize in a change in the size and relative stability of an obligor's current or projected revenue base, its operating requirements, its profitability or earnings, its cash flows or liquidity, or the size and maturity of its financial commitments. ESG risks influence the overview of a bank's credit strength as they affect the assessment of its asset quality, capital strength, profitability, liquidity, and funding. For example, climate change may undermine the repayment capacity of borrowers in weather-dependent or carbon-intensive sectors, reducing both asset quality and profitability of the bank.

The following are examples of ESG credit factors that can have positive as well as negative credit impacts, such as the reduction of social or environmental risks or the creation of earnings opportunities.

Examples of environmental credit factors are the following (illustrative and not exhaustive list):

- Greenhouse gas emission factors, including CO_2 emissions.
- Natural conditions factors, such as weather events.
- Other pollution factors, different from greenhouse gases.
- Other environmental factors, for example, water and land use and biodiversity.
- Environmental credit benefits, for example, factors that create revenue and earnings opportunities or reduce environmental risks.

Examples of social credit factors are the following (illustrative and not exhaustive list):

- Health and safety factors, for example, safety violations that lead to financial and reputational damage.
- Consumer-related factors, for example, mis-selling of products, related to environmental and social factors.
- Human capital management factors, for example, factors related to employee disputes and productivity.
- Social credit benefits, for example, factors that generate revenue and earnings opportunities or decrease social risks.

Examples of governance credit factors are the following (illustrative and not exhaustive list):

- Strategy, execution, and monitoring factors.
- Risk management and internal control factors.
- Transparency factors, including factors related to the quality of information disclosure.
- Board-related factors, including factors concerning the board's composition, independence, turnover, skill sets, gender, nationality, culture, and oversight of management.
- Other governance factors.

4.5   The Role of Technology in Overcoming ESG Data-Related Challenges

Initiatives to integrate ESG factors are, of course, related to ongoing digitization projects. How does technology help to solve the ESG data challenges for banks and investors? Technology plays a crucial role in empowering banks to overcome ESG data-related challenges and to align their investments with their ESG goals more effectively. The need to automate and to accelerate the response times is a priority in order to reduce the cost of customer services and to increase their quality. A further challenge in the near future will therefore be the convergence of these two project strands, which is essential to exploit their related synergies. This challenge is therefore a complex one because the availability of data is currently unsatisfactory. For example, the lack of comparability between databases provided by external parties is accentuated by the existence of numerous small firms, for which there are considerable information gaps in this regard. The insufficient availability of quantitative data, the inadequate consistency of rating methods by providers, and the lack of data standardization across markets are crucial challenges for ESG investing. In this context, Artificial intelligence (AI) technologies are increasingly used by financial institutions to improve their ESG performance and ability to promote ESG practices. Can AI offer innovative and transformative solutions to optimize supply chain systems as well as to face the complexities of regulatory compliance and to improve ESG assessment? To answer this question, it is important to be aware that AI and sustainability are two phenomena that have rapidly become closely interrelated. AI represents a very powerful knowledge tool and knowledge is the key factor in achieving all sustainability objectives. AI can be considered an accelerator for achieving sustainability objectives, a tool to facilitate and speed up reporting procedures and a technological solution to manage and reduce risks. In particular, the integration of AI within the banking sector's supply chain systems define a crucial path toward sustainable and equitable practices, allowing banks to assess and manage ESG risks across their interrelated systems. This includes making more sustainable investment decisions as well as contributing to a more environmentally and socially responsible financial ecosystem. In this sense, banks can use AI to promote ESG practices. With increasing pressure for organizations to adopt sustainability policies, the banking sector is looking for innovative ways to fulfill their ESG investment criteria. In this effort, AI emerges as an unexpected ally, offering ways of exploiting advanced analytics and machine learning (ML) to optimize supply chain ecosystems and to navigate the complexities of regulatory compliance on ESG assessment.

By automating tasks, identifying patterns, and making predictions, AI helps businesses to reduce their environmental impact, improve their social responsibility, and strengthen their governance. For example, AI is capable of evaluating the greenhouse gas (GHG) emissions from each supplier, choosing suppliers that satisfy minimal sustainability standards, minimizing waste and promoting fair working conditions and fostering economic growth. AI is applied to collect and analyze vast amounts of ESG data; for example, open source data and software tools can be used to meet the banks' urgent demand for transparent, consistent, and interoperable climate-relevant data within the ESG project data set. Offering a data-based approach that is supported by AI tools can enhance transparency in the supply chain and then help financial institutions to be ESG accountable. Since the lack of reliable and accessible data is the main barrier when assessing the cost of risk, banks can use shared and open source data to address this gap by integrating Artificial Intelligence (AI) in the context of financial technology (fintech).

Anyway, AI plays a pivotal role in some banking functions and processes. In Risk Assessment, AI can be used to assess a company's exposure to ESG risks, such as climate change, labor practices, and corporate governance. This information is crucial to develop strategies focused on mitigating risks and protecting a company's reputation. As concerns insurance companies, AI can support the development of new products and services that are more aligned with ESG goals. For example, AI can be used to assess the risk of natural disasters, which can help insurance companies to price their policies more precisely. Especially in the banking sector, AI can be used to identify sustainable investment opportunities, for example, by screening clients on the basis of ESG criteria (e.g. having low carbon emissions or strong labor practices) or supporting impact investing.[15] In this regard, the implementation of AI can improve the engagement of stakeholders and gather feedback on ESG initiatives from them. Finally, banks use AI to improve their lending practices. For example, AI assists banks in assessing a borrower's creditworthiness more accurately, which can help to reduce the risk of lending to borrowers who are unable to refund their loans.

---

[15] Impact investing is an investment activity in companies, organizations, and funds that operate with the aim of generating a positive social and environmental impact, which is measurable and compatible with an economic return. Impact investing seeks to achieve quantifiable social and environmental goals along with competitive financial returns. By allocating capital to companies that offer solutions to the world's most pressing problems, impact investors seek to generate positive change in society without sacrificing capital growth.

In this sense, AI emerges as a relevant tool, offering innovative but easy-to-use solutions for processing the vast and different data regarding of ESG metrics. In particular, AI is fully able to automate data collecting, organizing, and mapping key data from disparate IT systems, and then deploying them in a coherent and reliable ESG model. In this way, banks are able to meet regulatory requirements more efficiently to align with ESG criteria. As the future will move toward more data-heavy and complex technology, AI-driven analytics will become crucial to drive banks in managing their portfolios' social impact. Furthermore, where it is difficult to undertake an extensive due diligence before investing due to a lack of comprehensive data on sustainable investment options, AI can enable a fast access to relevant ESG data to produce sustainability analyses.

AI offers transformative innovations but it also poses significant challenges that require careful consideration. AI could address critical challenges while unlocking new opportunities for ESG practices in the banking sector and in this regard, it is important to be aware that this powerful tool has also potential disadvantages to be mitigated. In particular, we underline that the AI's potential introduces significant concerns, particularly concerning privacy and data integrity. AI models are built on data, and then if that data is biased, then the model will be prejudiced as well, leading AI to make unfair or discriminatory decisions. For example, an AI-focused lending model that is powered on biased data from the past could deny loans to certain people, even if they are just as creditworthy as other borrowers. In addition to data bias, since AI systems often collect and analyze large amounts of data, this system can upsurge privacy concerns. For example, an AI-based marketing system that tracks people's online activity could extract data about their personal preferences and purchasing behavior, which could then be used to target them with push advertising. From this perspective, the extensive data collection at the core of the AI-driven ESG analysis lifts privacy issues with the risk of algorithmic and/or human biases, further hindering the reliability and accountability of AI-generated information. Despite these challenges, AI's innovative capabilities promise improved accuracy over time. Anyway, the mentioned nature of AI in ESG assessment highlights the need for robust data transparency, accountability, and safety to ensure that the technology's advantages do not come at the expense of ethical considerations or data security.

There are other technologies that can help to solve the challenges related to the ESG data in banking activities. For instance, ESG data can be efficiently distributed in dedicated platforms. ESG data platforms can help investors in the investment decision-making process within the wide range of ESG data providers. In this regard, a fintech platform should permit investors to improve the portfolio data quality and to connect the portfolio data to multiple ESG providers in an optimized and efficient manner. In particular, more granular data sourcing and ESG data verification can be realized through the tokenization.[16] In this context, sourcing consistent and comparable data is often a challenge since the two main sources of information are from corporate reports and specialized third-party ESG ratings agencies. By using tokenization, that is, the process of representing real-world assets or data as digital tokens on a blockchain, investors can track the source of the ESG data and verify its authenticity across the overall value chain, facilitating data collection, particularly for smaller projects. Digital tokens can embed data on a blockchain in an efficient and traceable manner, allowing for greater transparency and ability to verify data even ex post throughout the value chain.

Another important aspect of transitioning toward more sustainable practices based on ESG data is being able to investigate the complex scenery of regulations through the establishment of a uniform framework for measuring the ESG compliance. The absence of universally adopted standards like the mandatory ones for financial reporting generates a significant difficulty in this regard. Financial institutions often rely on a mix of company disclosures, open channels, and external providers for ESG data which create an overall complexity. Hence, simplifying and growing up the capacity of the technological tools claim a systematic review of IT systems to incorporate ESG data effectively. This includes integrating and processing the sustainability Key Performance Indicators (KPIs) and Key Risk Indicators (KRIs) by means of smarter technologies, such as AI-based tools, that make easier the data management process for banks. Financial institutions can use AI algorithms to analyze vast amounts of data related

---

[16] Tokenization is the process of exchanging sensitive data for nonsensitive data called "tokens" that can be used in a database or internal system without bringing it into scope. Tokenization can be considered as a catalyst for ESG investing because investors can access more transparent investment opportunities, and small projects can reach a much wider set of potential investors.

to ESG factors and assess the risks associated with lending, investment, and underwriting activities. ESG-related data—like carbon emissions, diversity and inclusion metrics, or corporate governance practices—can efficiently be investigated during this process and financial institutions can use this analysis to score clients based on their ESG performance. These can then be used to create risk profiles and identify and mitigate potential ESG risks more successfully. For example, it could be identified as companies that join in greenwashing and/or companies that overstate their ESG credentials for marketing purposes. Overall, AI can be applied to automate compliance tasks (e.g. reviewing regulatory requirements and recognizing compliance gaps), producing more transparency and accountability. This can help banks to decrease costs and expand efficiency.

CHAPTER 5

Double Materiality and ESG for Banks

Abstract The chapter explains what exactly is double materiality, especially in the banking sector to achieve CSRD compliance. According to the CSRD, companies have to undertake a double materiality assessment to identify which sustainability matters are most material for an efficient allocation of the resources and for shaping bank strategy. The discussion starts from the difference between materiality and double materiality. Double materiality builds on the historical accounting and auditing convention of materiality and it takes this concept one step further by considering that companies are materially vulnerable to environment-related events and risks as well as they materially contribute to environmental issues. Hence, the ESG-related impacts on the bank can be material and the impacts of a bank on the ESG dimensions matter for materiality. These two perspectives have also been commonly referred to as "outside-in" materiality and "inside-out" materiality, respectively. On the one hand, banks have an impact on people and the environment (the inside-out view). On the other hand, sustainability-related developments and events create (new) risks and opportunities for banks (the outside-in view). Since double materiality is becoming a powerful concept to understand and put into practice the ESG strategy, the chapter describes double materiality, its rationales, proposals and challenges as well as the implications for implementing and realizing the ESG dimensions in banks. The chapter also shows some of the challenges in undertaking a double materiality assessment and it introduces some of the strategic implications for banks.

E. Menicucci, *ESG Integration in the Banking Sector*, https://doi.org/10.1007/978-3-031-81677-2_5

Keywords Materiality • Double materiality • Stakeholder engagement
• Banking sector

5.1 Materiality and Double Materiality

An important debate has started around the concept of double materiality,
reviewing, and expanding the existing accounting and auditing perspectives of materiality. The principle of materiality suggests that a firm's
accounting and reporting framework should reflect all the information
that could influence the decisions made by the users of the firm's financial
statements, such as stakeholders and particularly investors.[1]

Before delving into the idea of double materiality, it is important to
understand the concept of materiality itself in accounting and reporting.
How can we understand what exactly a material impact is? Materiality
refers to the relevance and/or the significance of information in decision-
making processes of companies. In sustainability reporting, materiality
helps firms to identify and prioritize the ESG dimensions that have the
major influence on corporate performance and stakeholders' demands.
According to this perspective, by focusing on material issues, banks assure
that their reporting is informative and that it contains the most important
sustainability concerns. In this regard, the International Accounting
Standards Board (IASB) had drafted the following definition, supported
by, among others, the European Commission: "information is material if
omitting, misstating or obscuring it could reasonably be expected to influence the decisions that the primary users of general purpose financial

---

[1] Materiality is a fundamental concept in financial reporting under IFRS Standards. See
IASB, IFRS Practice Statement 2: Making Materiality Judgements, Practice Statement 2024
Issued. IFRS Practice Statement 2: Making Materiality Judgements was issued in September
2017 for application from September 14, 2017. Other Standards have made minor conse-
quential amendments to IFRS Practice Statement 2: Making Materiality Judgements, including Amendments to References to the Conceptual Framework in IFRS Standards (issued
March 2018) and Definition of Material (Amendments to IAS 1 and IAS 8) issued October
2018. IFRS Practice Statement 2: Making Materiality Judgements (Practice Statement) provides companies with guidance on how to make materiality judgements when preparing their
general purpose financial statements in accordance with IFRS Standards. The need for materiality judgements is pervasive in the preparation of financial statements. IFRS Standards
require companies to make materiality judgements in decisions about recognition, measurement, presentation, and disclosure.

statements make on the basis of those financial statements".[2] In line with this perspective, the US Security and Exchange Commission (SEC) considers that "materiality concerns the significance of an item to users of a registrant's financial statements. A matter is 'material' if there is a substantial possibility that a reasonable person would consider it important." Hence, according to the US Securities and Exchange Commission (SEC), information on a company is material and it should be disclosed in financial statements if "a reasonable person" would consider it important.

Double materiality is an emerging concept that is developing rapidly, with potential implications for monetary and financial policies. Double materiality takes the concept of materiality one step further by considering that both the climate-related effects on the bank and the impacts of a bank on the dimensions of sustainability (e.g. societal and environmental impacts) can be material. Hence, double materiality extends the traditional concept of what accounting standards recognize as "material" by imprinting that ESG issues can affect a bank's financial performance (financial materiality). Double materiality builds on the historical principle of materiality in accounting and auditing and expands it by considering that companies are materially subjected to environment-related events and risks and they materially contribute to environmental deterioration at the same time. In other words, double materiality refers to a combined assessment of the vulnerability of the bank to environmental risks/events (the financial materiality) and the contribution of these companies to such risks/events (the environmental or impact materiality). The concept of dual materiality in the perspective of ESG refers to the analysis and evaluation of how environmental, social, and governance issues influence the financial performance of a bank (in the dimension of financial materiality) and, at the same time, how corporate activities impact on the same issues (in terms of external materiality or impact). Hence, this approach considers both the impacts of banking activities on the environment and society (external materiality) and the effects of ESG dimensions on bank performance (internal materiality). It is like having a big picture from two different visual angles since banks must consider the relevance of a sustainability matter from two perspectives. On the one hand, organizations have an impact on people and the environment (the inside-out view); on the other

---

[2] The IASB issued the Definition of "Material" (Amendments to IAS 1 and IAS 8) in October 2018 to clarify and align the definition of material.

hand, sustainability-related developments and events generate (new) risks and opportunities for organizations (the outside-in view).

This double-check idea has also been commonly recognized as "outside-in" materiality and "inside-out" materiality, respectively. The impact of banks on people and planet is outwards, while the sustainability and climate impact on the banks is inwards. Financial materiality (i.e. the outside-in perspective) arises from the influence of ESG factors on the bank's economic and financial activities throughout the entire value chain (both upstream and downstream) and then it influences the value (returns) of such activities. Hence, in the financial materiality concept, the focus is on how ESG aspects impact the banking business (e.g. bank's growth, performance, and cost of capital in the short, medium, and long term).[3] In other words, the concept underlines that banks should report simultaneously on sustainability matters that are both financially material in affecting corporate value and material in influencing the environment and society at large. As such, financial and non-financial institutions should disclose these impacts. For example, the EU's guidelines on reporting climate-related information have "a double materiality perspective" as they require companies to report on both financial materiality and impact materiality.

In particular, the principle of dual materiality encompasses two directions of analysis that can be summarized as follows:

1. From the outside-in perspective (in terms of financial materiality), the concept of double materiality incorporates the evaluation of how ESG factors can affect the financial and operational stability of an organization. For example, how climate change can develop a material risk for a firm causing a damage to production facilities or a disruption of the supply chain. In the same way, social issues and/or working practices fit into the concept of double materiality because they can impact on the firm's reputation and its performance.

2. From the inside-out perspective (in terms of external materiality or impact materiality), the concept of double materiality represents how corporate operations and policies can influence the ESG issues, which could in turn become financially material when this impact affects the returns of the corporate activities. In this analysis, we include, for example, the use of resources, waste management, the

---

[3] See EBA, EBA Report on management and supervision of ESG risks for credit institutions and investment firms, EBA/REP/2021/18, § 2.2, 41.

impact of greenhouse gas emissions, corporate carbon emissions that contribute to climate change, working conditions, and the contribution to the well-being of local communities, fair labor practices that can improve the social well-being of the surrounding environment.

According to the above considerations, double materiality entails that ESG factors can influence the long-term financial performance of a bank and should, therefore, be incorporated into the decision-making processes and the reporting for stakeholders. In this regard, banks have recognized that ESG risks and opportunities can have a significant impact on their long-term success and value. For instance, ineffective management can lead to higher operating costs, while good governance can improve corporate reputation and engage investors.

However, the ways in which double materiality should be implemented remain unclear. In this regard, monetary and financial authorities could realign their practices and policies with this concept, for instance by regarding the environmental impacts of their own operations and/or of the financial institutions they supervise. The concept of double materiality is progressively debated among authorities which are divided between those who oppose the concept and those who support it (e.g. in the European Union). Moreover, the debate exists between investors and regulators. In particular, the actual scope and application of the concept remain unclear and the potential implications of it could be subjected to various interpretations for monetary and financial policies. In order to clarify the existing idea of double materiality,[4] we can recognize different main approaches through which it can be inspected.

In conclusion, ESG dual materiality analysis offers a holistic view that allows banks to pass through a progressively complex and interconnected context. There is no doubt that double materiality today has become a key principle for banks that adopt a sustainability strategy and that intend to communicate to their stakeholders both how they secure their economic value from ESG risks, and how they are involved in reducing their negative influence on the environment and society. Furthermore, the approach of

---

[4]As Täger (2021) stated, the concept of double materiality still needs to be framed with contents. See Täger, M. (2021), "'Double materiality': What is it and why does it matter?", Commentary, April 21, Grantham Research Institute on Climate Change and the Environment.

double materiality helps banks to establish a more resilient and sustainable future, contributing to their long-term stability and growth of the planet. ESG dual materiality analysis can help banks to identify new business opportunities, for example, by developing new products or services that reply to environmental or social challenges, generating real value for both the corporate business and the society. In this regard, we believe that the developments of the ESG dual materiality are revolutionizing the business world and especially the financial industry. This approach of the ESG dual materiality has become increasingly relevant as the importance of sustainable and responsible investing increase, and simultaneously investors tend to pay growing attention to the financial performance of the banks in which they invest as well as to the impact of banking activities on the environment and society.

5.2   DOUBLE MATERIALITY ASSESSMENT ACCORDING TO CSRD

The CSRD significantly increases the number of disclosures that companies have to make about "material" sustainability matters in addition to the other disclosures. Specifically, companies that must report under CSRD have to undertake an assessment of the double materiality to identify which ESG matters are most material for the companies and their stakeholders. As specified in the previous paragraph, there are two types of materiality that firms have to consider when realizing their double materiality assessment according to the CSRD. The double materiality assessment determines the organization's sustainability reporting and it drives an efficient allocation of the resources needed to achieve CSRD compliance. Undertaking a "double materiality" assessment is mandatory under the Corporate Sustainability Reporting Directive (CSRD)[5] for companies that must report according to the CSRD. This regulation helps firms to

---

[5] On January 5, 2023, the Corporate Sustainability Reporting Directive (CSRD) entered into force. It modernizes and strengthens the rules concerning the social and environmental information that companies have to report. A key concept in CSRD is double materiality, which ensures comprehensive transparency and accountability in sustainability reporting. The CSRD requires businesses to assess and disclose sustainability matters from both the impact and the financial perspectives. The concept of double materiality is not only deemed within the context of the CSRD because other legislations address the topic of double materiality. For example, the Task Force on Climate-related Financial Disclosures (TCFD) recognizes the financial effects of climate change and promotes companies to evaluate and report

detect which ESG issues are material and should be included in their reporting. An identification of which topics are the most relevant (i.e. material) for a firm, and therefore must be included in its sustainability reporting, is an essential first step toward the CSRD-compliance. Such assessment identifies, in line with the CSRD-specific requirements, which aspects should be reported in a firm's sustainability reporting, and which ones can rightly be omitted. Making a double materiality assessment according to the CSRD entails some steps. Since banks can follow the double materiality assessment with a multi-step process, we explain the approach for each step providing insights from a practical perspective.

First of all, the bank should identify its relevant stakeholders and engage them to collect information about the bank's effects on the environment and society, as well as ESG risks and opportunities. Stakeholders are crucial for the double materiality assessment and the European Sustainability Reporting Standards (ESRS) in this regard introduce new considerations concerning the groups of stakeholders to involve, that is, the suppliers, partners, and various departments in the bank. Hence, who are the stakeholders that are affected by the bank? And which stakeholders can impact the bank? To answer to these questions, mapping stakeholders can be useful to identify which specific ones are directly involved in the materiality assessment process. The aim of stakeholder engagement is to recognize how people may be influenced by the bank. Moreover, through the stakeholder engagement, banks might get input and feedback on new and material sustainability issues to be included in their materiality assessment. In this perspective, under CSRD the role of stakeholder within the concerned process is modified. Now stakeholders are asked to identify the most significant impact of corporate activities on people and the environment as well as the most significant ESG risks and opportunities for the organization rather than the topics that they considered important.

Secondly, it is important to identify and define the relevant topics to be disclosed. Banks have to identify specific ESG matters. In particular, according to the ESRS, a bank should define a list of potentially relevant ESG matters and then determine which topics from the ESRS lists are relevant and material for the bank (e.g. sectors of activities, geographical areas of operation, and the entire value chain). Previous materiality assessments, internal documentation (e.g. impact and risk assessments), external

climate-related risks and opportunities. Similarly, the Global Reporting Initiative (GRI) acknowledges the concept of double materiality in its reporting framework.

documentation (e.g. sector reports, benchmarks, and ESG ratings), and insights from stakeholder engagement can be considered as sources of inputs. This step requires banks to understand what happens in the value chain and to recognize overall sustainable developments that can affect the business in various ways. Although the CSRD provides some guidelines for this analysis, finally the bank will have to determine if a matter is or not material. In this process, the concept of double materiality ensures that sustainability reporting focuses on the topics that are most relevant for the bank and its stakeholders. A strategy based on material topics realizes more transparency, contributes to better decision-making, and assures that time and resources are focused on those topics that better regard both the bank, its stakeholders, and society at large. Once the banks have identified the significant ESG topics for the organization, it is necessary to assess their impact, risks, and opportunities; this involves the consideration of both their significance in terms of environmental and social impact (impact materiality), as well as in terms of their financial impact (financial materiality). The definition of these issues in terms of bank's impacts, risks, and opportunities—as required by the ESRS—is the third step of the process of determining whether the ESG matters as identified in the previous step are indeed material (and should therefore be disclosed in the sustainability statements). This can be a challenging operation since the impacts related to any ESG matter on people and the environment can be positive or negative, potential or actual, and interrelated with the impacts from other issues. Moreover, these impacts, risks, and opportunities can verify in the short, medium, or long term and they can be reflected on future events and activities across the value chain.

Once ESG matters have been described in terms of impacts, risks, and opportunities, the next step is to quantify the identified impacts, risks, and opportunities. This step consists in measuring the effects of ESG risks and opportunities on the economic value of the bank, that is, the assessment of the financial effects that have not yet been included into the financial statements. Once again, this is a challenging exercise because it requires the comprehension of what is happening in the value chain, as well as insights into ESG developments that can affect banking processes. Moreover, organizations need to disclose how they manage these impacts, risks, and opportunities related to each topic. Going into details on these aspects allows to clarify their potential strategic implication. Input for these quantitative evaluations can be attained from engaging with stakeholders and experts, both internally and externally (e.g. through interviews, surveys, and workshops, etc.). The level of detail implied in this step facilitates a granular assessment to later determine which disclosure requirements are material according to the parameters defined in the ESRS. The assessment of risks and opportunities requires inputs from professional investigations that can help to identify events that might trigger a risk or opportunity (e.g. new regulations, increased public scrutiny, or changing stakeholder expectations regarding certain subjects). The assessment can support and ensure alignment with the broader enterprise risk management approach by assessing the magnitude of the financial effects (e.g. in terms of increase in R&D expenses, loss of revenue or increase in operational costs, etc.). Once all impacts, risks, and opportunities have been quantified, the bank can create separate rankings (high to low materiality score) for negative impacts, positive impacts, risks, and opportunities. By applying thresholds or cut-off points into the list, the values of impacts, risks, and opportunities can be split in material (top) and not material (bottom).

Finally, the CSRD requires that banks disclose precisely which actions they undertake to manage the environmental and societal impacts and the strategic implications for each sustainability matter that has been recognized as material. As a result, banks have to disclose the metrics and targets they have to apply for each ESG measure and the policies and action they will execute to achieve their goals. In addition, the CSRD requires banks to make disclosures about how they account for sustainability matters in their strategic planning processes.[6] This includes how material impacts, risks, and opportunities arise from the business model and how they require an adaptation of the market position and the value chain currently and in the future. Hence, banks have to take a longer-term perspective when they develop corporate strategy since the above described assessment's requirements require progressively to consider sustainability impacts. In this context, the disclosure of action plans requires banks to formulate credibly how they will guarantee that ESG matters are addressed in the banking functions and which parts of them are more involved in the organization.

---

[6] See Loizzo, T., Schimperna, F. (2022), Questioni di Economia e Finanza (Occasional Papers). ESG disclosure: regulatory framework and challenges for Italian banks, Banca d'Italia Eurosistema, Number 744 - December 2022.

5.3   APPROACHES TO DOUBLE MATERIALITY

The concept of double materiality can be investigated by means of different approaches. In this paragraph we illustrate three different rationales for interpreting double materiality despite an overall consensus around its definition. We inspect the rationales behind each approach, the possible theoretical and practical challenges, and the different potential implications for monetary and financial policies that could arise from the implementation of such settings. In this context, the adoption of a double materiality perspective remains a key operating question because the concept has the opportunity to interpret more comprehensively about the role of the financial system and banking industry in entirely addressing the ESG challenges.

The unsystematic perspective considers that the environmental, social, and governance impacts of corporate activities are relevant as they provide information on the risks faced by the company itself. This first approach inspects the double materiality at the level of the bank and it mainly refers to idiosyncratic/unsystematic risks.[7] For example, if a bank is exposed to clients that generate particularly damaging environmental implications, the financial institution should be aware that these firms' turnover, market value, and credit rating are likely to be negatively impacted by specific regulation. Under this approach, double materiality is regarded as a significant issue because environmental concerns could influence financial risks, for example, through negative effects on a company's reputation or legal liabilities.

This approach to double materiality is similar to that of dynamic financial materiality according to which the sustainability effects can become financially material over time. For instance, the European Commission states that financial and environmental-social materiality "already overlap in some cases and are increasingly likely to do so in the future. As markets and public policies evolve in response to climate change, the positive and/or negative impacts of a company on the climate will increasingly translate

---

[7] Unsystematic risk is the risk that is unique to a specific company or industry. It's also known as nonsystematic risk, specific risk, diversifiable risk, company-specific risk, or residual risk. In the context of an investment portfolio, unsystematic risk can be reduced through diversification, while the systematic risk is the risk that is inherent in the market. Unsystematic risk is a risk associated with a particular investment and it can be mitigated through diversification. Once it is diversified, investors are still subjected to market-wide systematic risk.

into business opportunities and/or risks that are financially material."[8] In this regard and in the perspective of policy proposals, regulators require that financial institutions report both the ESG impacts of their activities and the main financial risks that derive from such impacts. Results from the double materiality assessment can serve as a basis for dialogue with investors, creditors, or supervisors as well as for the potential challenges from them. This approach to double materiality in discussion outlines relevant challenges. The main one is that only the focus on financial materiality will likely not be sufficient to drive banks toward a longer-term horizon perspective: for example, a bank could rationally invest in short-term maturity activities and/or very liquid instruments that do not incorporate major financial risks in the short term but that are risky in the medium-long term.

A second perspective on double materiality is to consider it through the perspective of systematic risk, that is, without assuming that a bank's support to environmental decline is always reflected by its own vulnerability to future risks. A systemic risk perspective looks for reducing banks' support to negative environmental externalities, because of the systemic financial risks that could derive from them. This perspective considers double materiality at the financial system level and is focused on alleviating systematic or even systemic financial risks. This systematic perspective could even lead to a systemic issue. Actually, financial regulators and supervisors could consider that banks can accumulate future physical risks by financing polluting activities today. These risks could become systemic and irreversible, especially if certain limits are exceeded or if parties affected by environmental disasters ask for a considerable compensation (i.e. liability risks) so that these risks could become systemic. On the contrary, banks may contribute to the development of transition risks, which could also become systemic in certain scenarios (for instance, many banks that are exposed to future locked assets could imply fire sales if new sustainability regulations lead them to suddenly revalue the price of such assets).

This systemic risk perspective on double materiality also uprises significant challenges, for example, the circumstance that financial regulators could address challenging trade-offs between different time horizons and the thresholds through which they may want to value their objectives. For

---

[8] See Communication from the Commission, Guidelines on non-financial reporting: Supplement on reporting climate-related information (2019/C 209/01), Official Journal of the European Union, § 2.2.Materiality.

instance, it is possible to state that if regulators implement a "green assisting factor", they must adjust downward the risk-weighted assets (RWAs) for green assets, based on the rationale that this reduction mitigates physical risks at system level. On the contrary, they have to adjust RWAs upwards if the green activities are risky according to the Basel perspective (the time horizon may seem shorter than the horizon of climate and environmental-related risks). These considerations advise that regulators and supervisors could be less severe in considering short-term risks at an individual level under specific circumstances, when they foster longer-term stability of the financial system as a whole. In this context, it is also uncertain whether prudential policies would have relevant effects on the real economy, for example, whether these policies can indeed assist in decarbonizing the economy to enable the financial system in hedging against climate-related risks. For example, the possibility of applying a "non green sanctioning factor", that is, to growth the capital requirements required to regulated banks that fund carbon-intensive activities, can be an option although it is very debated among supervisors and academics. The two approaches above described focus differently on the confidence and solidity of the individual banks and of the entire banking system in achieving the mandate of the financial supervisors and the central banks in facing the environmental risks.

A third rationale regarding double materiality is the transformative approach. This perspective designs financial and corporate values and practices to make them more including of the different stakeholders' interests and more consistent with the activities that entail an ecological transition. This transformative approach tries to bypass the risk-based perspective and to review accounting and auditing regulations (e.g. the corporate reporting frameworks), in order to make them compliant with the ecological requirements. In order to face environmental challenges, actors in the financial context (e.g. central banks and bank supervisors) must actively contribute to the ecological transition, specifically by supporting the changing of accounting and auditing requirements regarding materiality. This perspective can therefore be interpreted as a robust concept of double materiality, in contrast to the unsystematic and systemic risk perspectives illustrated above, which we may consider a sort of feeble idea of double materiality.

From a reporting perspective, this approach underlines that disclosure on environmental effects is beneficial also because it apprises on prospective financial risks. Substantially, this approach tends to revise some characteristics of the financial system (e.g. the idea of materiality), to make them compliant for the aim of the structural transformation of our socio-economic systems and the ecological transition. Among these aspects that are required to be changed, we believe that the short-term perspective of the managerial incentives can involve choices that are not compatible with the achievement of long-term climate objectives, that in turn should be taken into account by future regulations. In this regard, financial and monetary policy implications of this approach are really wide. Financial regulators and authorities could proactively encourage financial practices that are more compliant with the ecological transition, for instance by penalizing loans for non-green activities.[9] Although some of these policies could be significant within the systemic risk perspective illustrated above, the majority of them could be better interpreted under the transformative perspective, possibly determining leading to trade-offs between different types of goals (i.e. traditional/primary and new/secondary objectives).

Similar to the prior perspectives on double materiality, the transformative approach outlines several challenges. While this may not be fundamental, one may ponder whether it is necessary to invent a new concept such as double materiality to promote new financial and corporate values. Hence, if some information is needed for non-financial reasons, it should be considered relevant as such and not supplanted by the concept of materiality.[10] According to an operational approach, operational compromises could arise for central banks could face a trade-off between the willingness to proactively incentive the ecological transition and the need to realize their traditional/primary mandates. For example, where climatic and environmental policies had an inflationary impact, a growth of interest rates issued by central banks would impact the achievement of environmental goals if related investments require a higher initial capital. On the contrary, seeking to reach price stability in the short term and at the cost of the ecological transition could compromise environmental and financial stability in the long term.[11]

[9] See Demekas, D.G., Grippa, P. (2021), Financial Regulation, Climate Change, and the Transition to a Low-Carbon Economy: A Survey of the Issues. International Monetary Fund.

[10] See Katz, D.A., McIntosh, L.A. (2021), "Corporate Governance Update: "Materiality" in America and Abroad". Harvard Law School Forum of Corporate Governance.

[11] See Bolton, P., Després, M., Pereira da Silva, L.A., Samama, F., Svartzman, R. (2020), "'Green Swans': central banks in the age of climate-related risks", Banque de France Bulletin, 229/8 (May–June).

5.4   THE CHALLENGE OF DOUBLE MATERIALITY
IN THE BANKING SECTOR

The banking industry is progressively focusing on how to best recognize and evaluate the material risks and their impacts on banking operations as well as on value chains, in line with the developing regulatory requirements concerning environmental, social, and governance (ESG) disclosure and in order to react to pressure from a large range of stakeholders. This is a very different challenge from previous ESG materiality approaches, which is mainly focused on financial risks related to the banking business alone. In this sense, the new concept of double materiality implies that banks must assess both the internal banking activities and the effects of the behavior of clients with regard to ESG factors. It is evident that, in line with this concept, ESG risks must be interpreted under a double point of view, that is, that proposed by EBA as an outside-in and an inside-out perspective. Hence, the relationship between the inside-out and outside-in perspectives is explained by the concept of double materiality, which is divided into financial materiality and environmental or impact materiality. According to the first perspective, banks can be affected by ESG risks through their counterparties and invested assets, but at the same time, they may influence the ESG dimensions (inside-out perspective). Even if both perspectives are important, the inside-out are becoming much more relevant. The financial materiality can be inspecting by considering the effects on the bank's economic and financial activities. On the contrary, the environmental or impact materiality refers to the influence of the bank's economic and financial activities on ESG dimensions themselves. For banks, understanding the concept of double materiality and reporting for it is crucial for sustainable decision-making and effective disclosure on ESG issues. Although every bank's double materiality approach should be tailored to its specific financial context, there are some general best practices that banks should consider when dealing with materiality and double materiality considerations in strategic decision-making.

Starting from the materiality principle, banks must turn toward the double materiality perspective. Certainly, materiality is a powerful concept to understand and to put into practice for boards, corporate executives, and sustainability practitioners. Its assessment must be projected to help a bank in organizing and prioritizing its material issues to drive its sustainability turnaround toward a specific ESG strategy. Differently, to perform a strategic double materiality assessment by effectively managing

ESG-related risks and its impacts on banking functions, it is important to assure a structured approach for identifying which environmental and social issues are most significant and challenging for the business and its stakeholders. This approach can be considered the most powerful for a banking business to better understand and implement double materiality (in particular, prioritize the ESG issues and decide how to allocate resources to enhance them). When strategic decisions that consider double materiality are made in banks, business leaders should evaluate the long-term implications of the business options and practices, and should try to anticipate and foresee how they will affect the bank's financial and ESG performance. Many sustainability reporting standards (i.e. the ESRS) require companies to disclose information about how they evaluate double materiality. Hence, it is important for banks to adhere to disclosure regulations, including new requirements, rules, and guidelines, especially those mandatory issued by European regulators, to mitigate potential legal liabilities. As concerns the practical implementation of double materiality, the process needs to be planned, scoped, temporized, socialized, and budgeted in the context of a bank's ESG reporting requirements. From a prudential perspective, ESG risks for banks are defined as the negative materialization of ESG factors through their counterparties or invested assets. Hence, banks can be impacted by (outside-in perspective) ESG risks through their counterparties and invested assets, as these may be affected by (outside-in perspective) or have an impact on (inside-out perspective) ESG factors. Both of these perspectives should be considered when assessing ESG risks. By proactively addressing these emerging risks and then by prioritizing appropriate related actions, a bank can materialize its commitment to ESG practices while bridging reporting requirements. How can banks effectively capture and evaluate ESG risks and their impacts? To realize an effective assessment through the lens of double materiality, the subsequent guidelines should be followed.

- Implement internal capacity and awareness. Building firstly a shared comprehension of double materiality is fundamental to make the assessment of the ESG strategies and their disclosure effective. In this regard, internal human rights, environmental, and social teams may be more equipped and proper for evaluating outward impacts, while business risk teams may not be accustomed to evaluate the risks for people and planet (i.e. environmental and social, respectively).

- Align with the results of the materiality assessment criteria in terms of impacts. When conducting double materiality assessments, the deep dive move from a stakeholder perception-based approach toward an approach based on impacts and specifically focused on assessing their likelihood and severity.
- Assure substantial stakeholders' engagement. Engaging stakeholders to recognize inward and outward ESG risks for the business should imply the involvement of different rightsholders or their representatives, who may be actually or potentially impacted by risks correlated to the business. It is important to include in the inspection critical entries as well as those closer to the banking business, in order to gain valuable insights to drive subsequent actions.
- Apply various methods to quantify impacts. A high-level double materiality assessment may find out wide banking risks and opportunities at corporate level. However, the capture of inward and outward impacts will be meaningful because it requires making deeper dives, for example, assessments to be conducted at portfolio- and supply chain-levels. This approach also implies to consider multiple data sources of information for gaining further insights into the impacts of ESG risks (ESG indices, benchmarks, and reports) over the business risk. In this perspective, to better inform about the assessments, the adoption of a forward-looking approach is crucial for a useful disclosure about emerging trends and scenarios that can change possible risks and impacts related to ESG factors. The credit rating model therefore should be able to identify, weigh, and understand the "double materiality" of the various relevant ESG variables to be considered adequate in terms of predictive effectiveness and range of the forecast horizon. These ESG variables refer to the impact of ESG factors on competitive positioning, development prospects and performance of the bank (financial materiality), and the impact of the bank's policies on ESG topics, in particular, society and the environment (impact materiality). For example, double materiality in the banking industry involves how banks revise their lending practices and loan portfolios to be secure that they not fund (even indirectly) business activities that are dangerous for the environment. These decisions can have a positive environmental effect but they also affect the bank's financial performance and its lending risk.

5.5 STAKEHOLDER ENGAGEMENT TO ENHANCE THE ESG STRATEGY THROUGHOUT THE MATERIALITY PROCESS

The process of the materiality's assessment is a relevant and efficient opportunity to engage stakeholders in implementing and updating the ESG strategy. Within the newly adopted EU Corporate Sustainability Reporting Directive (CSRD), the assessment of double materiality (which is mandatory from 2024 to all large European Union companies listed on stock markets or not as well as to non-EU companies with substantial activity in the EU with a turnover over Euro150 million euro in the EU) especially requires an active deep dive on the concerned stakeholders to identify and evaluate the current and potential effects in building the ESG strategy. Hence, the banks must take the opportunity for a productive dialogue with stakeholders as a step of the implementing process of its ESG strategy. The involvement of the stakeholders in the ESG materiality assessment start by engaging with internal and external stakeholders exactly, considering their inputs regarding the environmental impacts related to climate change. The inspection of the feedback returned from stakeholders and the assessment of the significance of the potential identified issues enable a bank to make prudently informed decisions about its ESG strategy and reporting.

In the ever-evolving banking context, environmental, social, and governance (ESG) factors have become essential in designing the banking forward-thinking activities. In particular, the identification and the involvement of the pertinent stakeholders become one pivotal variable for realizing strong ESG materiality assessments while the banks try both to align their actions with sustainability goals and to meet the expectations of overall stakeholders (especially, investors, regulators, etc.). At the core of the ESG materiality assessment there are the stakeholders who are significant for the final implementation of the ESG strategy. This approach guarantees that the bank's ESG strategy aligns with the most relevant and appropriate topics with the aim of maximizing opportunities for sustainable and long-term growth and minimizing the correlated risks. In this regard, we believe that the stakeholder engagement is a fundamental tool for the success of the ESG strategy and it represents a best practice solution to ensure ESG efforts drive meaningful contributions in banking operations. Any successful ESG strategy starts from a materiality assessment that is configured as a process taking various forms (including surveys, interviews, and ongoing dialogues with earlier recognized stakeholders). The engagement of stakeholders is crucial in the

assessment's process because material ESG issues are detected and prioritized by means of their input, concerns, and perspectives. These considerations often mirror the most relevant issues for a bank's operations because
they have the best ability to produce wide impacts. Besides sustaining the
development of a centered and inclusive ESG strategy, engaging a large
panel of stakeholders in the materiality assessment can help the bank to
better identify and address operational ESG factors that directly impact on
them or on their stakeholder's collectivity. Usually, banks drop knowledge
and awareness about these aspect as well as about the risks and the severity
of their impacts; hence, banks drive direct efforts to where they are most
needed through gathering insights from those who are most affected (i.e.
stakeholders). Stakeholders can identify with investors, employees, customers, suppliers, the local community, or substantially with any person
who has a gained interest in understanding the bank's activities and their
impact. Although it is expected that the interests of various stakeholders
may differ, the overall perspective of all stakeholders is crucial for the
implementation of a robust ESG assessment's process. For instance, investors may be mainly interested in governance themes and financial performance, while employees should give a major priority to job security and
career advancement opportunities, and local communities tend to give
higher importance to climate issues and environmental impacts. The real
challenge for leveraging on stakeholders' knowledge is the identification
of the proper stakeholders among a wide range of them. Then, by leveraging stakeholder knowledge, companies can not only save time and
resources, but also ensure their ESG strategies and underlying actions are
aligned with the most material issues for the company and its stakeholders.
Anyway, there is no one-size-fits-all approach for identifying stakeholders
because each bank is unique and the materiality assessment process for
everyone should be tailored to its specific resources, requirements, objectives, and desires concerning sustainability.

Therefore, engaging stakeholders is vital for banks to implement a solid
ESG strategy, to meet ESG commitments, and to rank their impact.
However, effectively involving stakeholders in achieving these ESG objectives implies other challenges. How can banks engage their stakeholders
while carrying out their materiality assessment? Can stakeholder engagement accelerate ESG performance? We can also consider how different
materiality assessments can promote and enhance the process of stakeholder engagement. More specifically, there can't be a successful ESG
strategy without a materiality assessment and there can't be an effective
assessment without involving the proper stakeholders. We inspect more

practical insights on stakeholder engagement to find how to make the best out of it. In other words, stakeholder engagement is a key for developing a bank's ESG performance and for building a solid ESG strategy to consider stakeholder engagement in ESG materiality. Stakeholder engagement is the process by which banks involve their stakeholders toward a wished outcome that can range from making additional changes to operational processes to making structural improvements in the organizational strategy. The engagement itself can appear differently such as having discussions with local collectivity to review infrastructures, setting goals for senior management for a better ESG performance, developing partnerships with other players in the sector to progress in circular economy issues, and any other types of exchanges with the different stakeholders. Banks are constantly driven by and concerned with their stakeholders' expectations. We refer, for example, to customers' expectations about products'/services' performance, employees' needs with regard to working conditions, and suppliers on enduring partnerships. Considering these expectancies, stakeholders play a pivotal role in modeling the strategy in the opportune way. Therefore, stakeholders must be identified, prioritized, and checked for the design of the strategy, in such a way that it reflects stakeholders' expectations and inputs. Stakeholders are those parties who affect and/or are affected by a bank's decisions and actions. They are grouped into two clusters: internal stakeholders that can directly impact the bank from the inside (employees, senior management, and board members) and external stakeholders that affect the bank from the outside (investors, customers, suppliers, local communities, public organizations, NGOs, research partners, etc.).

Therefore, stakeholder engagement allows banks to increase their impacts on society in addition to building a strong ESG strategy and meeting ESG commitments. The impacts can be materialized in terms both of speed (accelerating the changing and the operating) and magnitude (enhancing the positive impacts). Once the direction of the bank's strategy has been set, the engagement of stakeholders should continue to be a priority for the execution of the strategy and the realization of the ESG commitments. Having involved stakeholders from the beginning ensures that the strategy already mirrors their expectations and then they will be more inclined to be available once they are asked. The stakeholder engagement regards, for example, the collaborating with partners to decrease the effects on the whole value chain, training employees on ESG themes, and requirement of investments from the executives, etc. Initially, banks should well define objectives and conduct an initial screening of the possible impacts to guide the preliminary stakeholder selection and mapping process. Although a normative guide or a standard guideline for the selection of the stakeholders would be difficult to find, there are a number of best practices that banks can use for their stakeholder identification process to gain their strategic objectives. For example, when a bank identifies the necessity for a focused regulatory compliance, it should engage regulatory authorities, legal teams, and sector associations; otherwise, when a bank may want to safeguard itself against future environmental-related risks, it should exploit the competence of its investors, employees, supply chain partners, and even external environmental specialists.

An initial mapping process to design a complete panel of stakeholders ensures the bridging of the gaps in the materiality assessment. Hence, banks should recognize and group various internal and external stakeholders who are (or may be) interested and affected by the company's operations for comprehending their degree of relevance, significance, influence, and engagement level. The use of tools (e.g. matrices or power-interest grids) can aid the refining process to better guide the subsequent engagement efforts. To assure solid, unbiased, and transparent outputs, banks must engage different stakeholders covering the whole value chain and representing personal differences (e.g. gender, ethnicity, etc.). The quality of the assessment will also depend on the inclusion of a wide variety of perspectives, in addition to the most obvious or powerful stakeholders. Balancing the number and types of stakeholders is pivotal for the success of a materiality assessment. Involving too few stakeholders decreases the assessment's scope, damages the bank's reputation and relationships, and is translated into a limited and probably inaccurate viewpoint of the bank's ESG issues. On the other hand, engaging a large number of stakeholders can make the assessment complicating, costly, and time-consuming. It is important to acknowledge that when a business evolves and changes, the importance of certain stakeholders will develop together. Banks should regularly revise and adjust their approach to stakeholders' identification and selection to mirror these evolutions.

Engaging stakeholders is indeed a multiform process. In this sense, for maximizing the value of the engagement activities, some aspects are crucial, such as nurturing stewardship and trust, educating, dedicating adequate time, and recognizing and mitigating engagement risks. Before starting the materiality process, banks can enhance the probability of receiving detailed inputs by delivering wide information to their stakeholders regarding ESG (issues, challenges, risks, frameworks) to guarantee a shared comprehension. This ensures that all stakeholders have a strong knowledge of the request and a solid ability to provide informed input that finally can lead to high value goals. Stakeholders may sometimes be reluctant to share their perspectives. Hence, the banks need to cultivate trust and stewardship to facilitate open dialogue with stakeholders and encourage them to express their real viewpoints.

ESG materiality assessment and stakeholder engagement are closely connected and dynamically weaved together. They are constantly developing processes that are increasingly becoming crucial for the bank's long-term success. ESG materiality assessment can be considered a guide for banks for navigating the multifaceted landscape of ESG issues and aligning their activities with evolving requirements. Engaging the proper stakeholders encourages both to identify and prioritize material ESG topics and to promote an inclusive and transparent approach, fostering the credibility of a bank's ESG efforts. The key point is to be determined in realizing the engagement by clearly underlining the importance of bridging the knowledge and competence gaps and proactively addressing the barriers. By building robust and stakeholder-informed ESG strategies, banks can ensure the direction toward ambitious, relevant, and focused efforts that favor meaningful changes. Stakeholder engagement must develop together with the change of market conditions and the development of regulatory conditions, business sector, stakeholder expectations, and the introduction of new ESG requirements.

From a practical perspective, banks can tailor their approach to meet these specific objectives and to support developing priorities. In this regard, we illustrate the following three paths for successfully engaging stakeholders and for enhancing the ESG performance through the process of a materiality assessment.

1. Comprehend who the bank's stakeholders are. As a starting point, it is essential to understand the bank's stakeholders and how they affect and are affected by the banking activities. For a more incisive approach, stakeholders should also be prioritized based on different variables (e.g. applying the stakeholder salience model).[12] Based on

---

[12] The Stakeholder Salience Model is a strategic tool for segmenting stakeholders. The Model suggests a new way to identify stakeholders based on three concepts: power to influence the company, how valid their connection to the company is, and how urgent their claim on the company is. The ranking and the assessment of all stakeholders are based on three attributes: power (the ability to use force, money, or social standing to get what you want),

the potential ESG risks and opportunities, a bank can identify different stakeholder groups to be involved in the assessment including customers, employees, investors, local communities, public organizations, suppliers, and research partners. The banks also consider the criticalities of its business environment and can decide to involve internal and external stakeholders across the different business areas in which the bank operates.

2. Involve the stakeholders in implementing and evolving the ESG strategy. Materiality assessment requires an active interrelation with stakeholders to identify and prioritize issues that are material for a bank to build and update the ESG strategy. For instance, a bank can exhort the key stakeholders to take part in the materiality assessment to update its ESG strategy. Best practices used in this regard can include interviews, workshops, desk research, and surveys. The bank can combine different research methods to exploit their several advantages. For example, workshops with internal stakeholders are useful to share different viewpoints, interactions with external stakeholder panels can facilitate best practice to follow, interviews assist in exploring specific topics in-depth, and desk research helps to get an overall organizational perspective.

3. Develop a specific approach to every stakeholder group. Based on the feedback received during the stakeholder engagement process, it is important to apply a specific approach for each group of stakeholders to improve ESG performance. Each approach includes some operational activities although wide differences can upsurge based on the stakeholders concerned and the results of the exercise. What other activities in this regard can be a good idea? Setting and reviewing quantified targets; linking the objectives to incentives where appropriate; supporting the stakeholders in their activities and monitoring performance where it is necessary. Encouraging stakeholders to provide feedback and future possible options for win-win strategies; inviting stakeholders to activate workshops for recognizing real solutions for the relevant topics. Finally, taking action by building industry coalitions with peers and business partners in the reference context or fulfilling business reviews with suppliers/customers as well as educating to bridge the identified knowledge gaps.

legitimacy (when people think an action is right and proper based on social norms and beliefs), and urgency (how quickly a stakeholder needs attention. This depends on time and how important the claim is. The Stakeholder Salience Model involves an overview and a prioritization of all stakeholders as part of stakeholder analysis, management, and engagement.

INDEX

© The Author(s), under exclusive license to Springer Nature
Switzerland AG 2025
E. Menicucci, *ESG Integration in the Banking Sector*,
https://doi.org/10.1007/978-3-031-81677-2